I0119415

THE

ZOOLOGY

OF

THE VOYAGE OF H.M.S. BEAGLE,

UNDER THE COMMAND OF CAPTAIN FITZROY, R.N.

DURING THE YEARS

1832 TO 1836.

PUBLISHED WITH THE APPROVAL OF
THE LORDS COMMISSIONERS OF HER MAJESTY'S TREASURY.

Edited and Superintended by

CHARLES DARWIN, ESQ. M.A. F.R.S. F.G.S., Etc.

NATURALIST TO THE EXPEDITION.

PART IV.

FISH,

BY

THE REV. LEONARD JENYNS, M.A., F.L.S., &c.

Copyright © 2018 Read Books Ltd.
This book is copyright and may not be
reproduced or copied in any way without
the express permission of the publisher in writing

British Library Cataloguing-in-Publication Data
A catalogue record for this book is available from
the British Library

INTRODUCTION.

The number of species of Fish described or noticed in the following Part of the Zoology of the Beagle, amount to 137. It is right to observe that, judging from Mr. Darwin's manuscript notes, relating to what he obtained in this department, this is probably not more than half the entire number which he collected. Unfortunately a large portion of the valuable collection sent home by him arrived in this country in too bad condition for examination, and was necessarily rejected.

The localities visited by Mr. Darwin, and at every one of which more or fewer species of fish were obtained, were the Cape Verde Islands,—the coast of Brazil, including the mouth of the Plata, together with several inland rivers and streams in that district,—the coasts of Patagonia, and the Santa Cruz river,— Tierra del Fuego and the Falkland Islands,—the Archipelago of Chiloe,—the coasts of Chile and Peru,—the Galapagos Archipelago,—Tahiti,—New Zealand, King George's Sound in Australia,—and, lastly, the Keeling Islands in the Indian Ocean. The great bulk of the species, however, are from the coasts, east and west, of South America.

The particular locality assigned to each species respectively in the following work may be relied upon as correct; pains having been taken by Mr. Darwin to affix a small ticket of tin, with a number stamped upon it, to each specimen, and to enter a note immediately in the manuscript catalogue, having the same number attached. In only three or four instances these tickets were found wanting, on the arrival of the collection in this country.

A considerable portion of the species examined and described are new to science, especially of those collected in South America, and the adjoining Islands and Archipelagos. The new ones are supposed to amount to seventy-five at least, constituting more than half the entire number; and amongst these are apparently seven new genera.

a

INTRODUCTION.

It may be interesting to state more particularly from what localities the new species principally come, and what proportion they bear to the *entire* number brought from each of those localities. Thus from Brazil *about half* are considered new ;—from Patagonia *at least half;*—from Tierra del Fuego, the Falkland Islands, and the Galapagos Archipelago, *all are new,* without exception ; and *nearly all* from Chiloe, and the coasts of Chile and Peru. Of the species brought from Tahiti, New Holland, and the Indian Ocean, not above *one-fourth* are new. This might have been anticipated from the better knowledge which we have of the Ichthyology of that quarter of the globe, than of South America.

It is much to be regretted that the portion of the collection which has been lost to science, was obtained in localities most abounding in novelties, judging from that portion of it which has been saved. Thus, not above five or six species will be found noticed in the following work, from Tierra del Fuego, where Mr. Darwin took especial pains to collect all he could, and, judging from his manuscript catalogue, he must probably have obtained between thirty and forty. From the Falkland Islands again, there have been only saved two out of fifteen or sixteen,—from the coasts of Chile and Peru, not half the entire number obtained, and not above half from the coasts of Patagonia.

There is also described not above half the species brought from King George's Sound, and the Keeling Islands ; but as the Indian and Australian species, or at least the former, have been more frequently brought to Europe than the South American, they are less to be regretted than these last.

It is fortunate that *the whole* of the species obtained by Mr. Darwin in the Galapagos Archipelago, amounting to fifteen, have been preserved, and are described in the following pages.

It may now be useful to mention, to what groups principally—first, the entire number of described species belong, and, secondly, that portion of them which are considered new. Both these points will be best judged of from the following table, in which the whole collection is parcelled out according to the families.

ACANTHOPTERYGII.

PERCIDÆ.	Entire No. of species 18 whereof new 11					Brought up	.	.	. 45		22
MULLIDÆ	.	.	.	.	. 3	SCOMBRIDÆ	.	.	. 7	.	. 3
TRIGLIDÆ	.	.	.	. 3	. . 1	TEUTHYDIDÆ	.	.	. 2		
COTTIDÆ	.	.	.	. 2	. . 2	ATHERINIDÆ	.	.	. 3	.	. 2
SCORPÆNIDÆ	.	.	.	. 4	. . 2	MUGILIDÆ	.	.	. 3		
SCIÆNIDÆ	.	.	.	. 10	. . 5	BLENNIDÆ	.	.	. 11	.	. 7
SPARIDÆ	.	.	.	. 1	. . 1	GOBIDÆ	.	.	. 3	.	. 2
MÆNIDÆ	.	.	.	. 2		LABRIDÆ	.	.	. 7	.	. 5
CHÆTODONTIDÆ	.	.	.	. 2		LOPHIDÆ	.	.	. 1		
				45	22	TOTAL	.	.	. 82	TOTAL, NEW	41

INTRODUCTION.

MALACOPTERYGII.

Silurid. Entire No. of species 3 whereof new 2	Brought up . . . 30		21
Cyprinidæ . . . . . 7 . . . 6	Cyclopterid . . . . 2 . . . 2
Esocidæ 1	Echeneidid 1
Salmonidæ . . . . . 8 . . . 7	Anguillidæ . . . . 6 . . . 2
Clupeidæ . . . . . 5 . . . 5		[perhaps more.]
Pleuronectidæ . . . . 6 . . . 1	
[probably more.]	
30 21	Total . . . . 39 Total, new 25

LOPHOBRANCHII.

Syngnathidæ. Entire No. of species . . 3 whereof new . . . 3

PLECTOGNATHI.

Tetrodontidæ. Entire No. of species. . 7 whereof new . . 4
Balistidæ 5 1

Total 12   Total, new . . 5

CYCLOSTOMI.

Petromyzonidæ. Entire No. of species . . 1 whereof new . . . 1

TOTAL IN THE SEVERAL ORDERS.

Acanthopterygii. Entire No. of species . . 82	whereof new	. 41
Malacopterygii 39	. . . .	. 25
Lophobranchii 3	. . . .	. 3
Plectognathi 12	. . . .	. 5
Cyclostomi 1	. . . .	. 1
Grand Total 137	Grand Total, new	. 75

It appears from the above table that of the entire number of species, three-fifths belong to the Acanthopterygian fishes,—rather more than one-fourth to the Malacopterygian,—and about one-eighth to the remaining orders united.

In the Acanthopterygians, the *new* species amount to one-half; in the Malacopterygians, to about two-thirds; in the remaining orders together, to rather more than one-half.

Looking, therefore, to the entire number of species described, the Acanthopterygians prevail; and it is in the same order that there are most new ones: but looking to the proportion, which in each order the new ones bear to the entire number, it is among the Malacopterygians that this proportion will be found highest.

Restricting our view, it will be also seen, in the Malacopterygians, that the new species are relatively most numerous in the fresh-water groups, such as the *Siluridæ*, the *Cyprinidæ*, and *Salmonidæ*, in which three families taken together, they amount to five-sixths of the whole. The *Clupeidæ* are an exception, in which all the species are apparently new.

All the species described, belonging to the three families above mentioned, in which there are so many new, viz. the *Siluridæ*, the *Cyprinidæ*, and *Salmonidæ*, are from South America, and the Falkland Islands, excepting one from New Zealand.

Of the remaining fresh-water fishes in the collection, three out of five are presumed to be new. One of these is a species of *Perca*, from the Santa Cruz river, in South Patagonia; the second is a species of *Dules*, from the river Matavai, in Tahiti; the third a species of *Atherina*, from Valparaiso. Perhaps, however, this last is not strictly an inland species.

The entire number of fresh-water species in the collection is twenty-three, and the entire number of new ones amongst these is eighteen. The large proportion of these latter is a circumstance in confirmation of a remark which Cuvier has somewhere made, that the fresh-water fishes of foreign countries are much less known and understood than those found on the coasts. It may serve also as a hint to future travellers.

The seven new genera in the collection belong—one to the *Sciænidæ*, from the Galapagos Archipelago;—one to the *Scombridæ*, from North Patagonia;—three to the *Blennidæ*, whereof one is from the Archipelago of Chiloe, the second from the Falkland Islands, and the third from New Zealand;—one to the *Cyprinidæ*, embracing three species, from South Patagonia, Tierra del Fuego, and New Zealand; and, lastly, one to the *Salmonidæ*, embracing two species from the Falkland Islands and Tierra del Fuego respectively.

It has been already mentioned, that all the species obtained by Mr. Darwin in the Galapagos Archipelago have been preserved. As they are likewise all new, and those islands appear to have been scarcely visited by any naturalist previously, it may be interesting to enumerate the several genera to which they belong, and the number of species in each genus respectively.

SERRANUS .... 3 species.	Fam. PERCIDÆ.	
PRIONOTUS .... 1 ,,	—— TRIGLIDÆ.	
SCORPÆNA .... 1 ,,	—— SCORPÆNIDÆ.	
PRIONODES *N.G.* 1 ,,		
PRISTIPOMA .... 1 ,,	—— SCIÆNIDÆ.	ACANTHOPTERYGII.
LATILUS ...... 1 ,,		
CHRYSOPHRYS .. 1 ,,	—— SPARIDÆ.	
GOBIUS ...... 1 ,,	—— GOBIDÆ.	
COSSYPHUS .... 1 ,,	—— LABRIDÆ.	
GOBIESOX .... 1 ,,	—— CYCLOPTERIDÆ.	MALACOPTERYGII.
MURÆNA ...... 1 ,,	—— ANGUILLIDÆ.	
TETRODON .... 2 ,,	—— TETRODONTIDÆ.	PLECTOGNATHI.

15

In making the foregoing estimates, as regards the number of new species brought home by Mr. Darwin, I have been guided almost entirely by my own judgment. The difficulty, however, of ascertaining, in a miscellaneous collection of this nature, brought from various localities, what *are* really new to science, is very great; and this difficulty is much increased, where an author is situate apart from large public museums to which he might have recourse for comparison. Possibly, therefore, some of those described as new in the following work, may not be so in reality; and, in one instance, as mentioned in the Appendix, this is known to be the case. My excuse, however, must rest upon what has been just stated. It is hoped that caution has been generally shown, at least in regard to specimens not in a good state of preservation; and, in several such cases, in which an accurate description was hardly practicable,—though they could not be referred to any known species,—they are not positively declared new, nor any names imposed upon them whatever.

I have, of course, consulted throughout the invaluable volumes of Cuvier and Valenciennes, so far as they have yet advanced in the subject; and in them it will be found that a few species, brought by Mr. Darwin from South America, and still but little known, had nevertheless been previously obtained from the same country by M. Gay. The zoological atlasses of the three great French voyages by Freycinet, Duperrey and D'Urville have been also carefully looked through; and, in regard particularly to the fish of South America, the works of Humboldt, Spix and Agassiz, and the more recent one, now in course of publication, by M. D'Orbigny.

There is an equal difficulty felt by every naturalist at the present day, in distinguishing species from varieties. And in the case of Fish, residing in a peculiar element, and so much removed from our observation,—we are almost at a loss to know, at present, to what extent their characters may be modified by local and accidental causes, or how far we may trust a different geographical position for giving permanence and value to a slight modification of form different from what occurs in the species of our own seas. Still less easy is it to determine the true importance of characters, in instances in which it is only permitted to see a single specimen of the kind, or, at most, very few individuals.

Many mistakes, therefore, are liable to occur, in a work of this nature, arising from the above sources. The only way to prevent their creating any permanent confusion in the science, is to describe all species of which the least doubt is entertained, in such detail, and with such accuracy, that they may not fail of being recognized by any observer, to whom they may occur a second time. They will not then *continue to hold a false position* in the system, as *spurious*

2

species. They may not be new, or they may not be species at all,—but they will be *known;* and any mistake which has been committed will be at once rectified,—any new name which has been wrongly imposed, immediately degraded to a synonym.

Accordingly I have been careful in this respect; and I have in some instances, given full descriptions, even of species which are certainly not new, but which I did not find described by previous authors with all the detail that was requisite for completely identifying them ; or, leaving out what they have noticed, I have added such characters as they have omitted. My main object has been to render all the species, whether rightly named or not, easily recognizable ; and, however little the science may be advanced by what is brought forward, to make that advance, so far as it goes, sure.

The method of description, and the mode of computing the fin-ray formula, will be found conformable to the plan adopted in the "Histoire des Poissons" of Cuvier and Valenciennes ; a work which, in so many respects, must always serve as a model to labourers in this department of zoology.

The colours, in the great majority of instances, were, fortunately, noticed by Mr. Darwin in the recent state. The nomenclature employed by him for the purpose is that of Patrick Syme; and he informs me, that a comparison was always made with the book in hand, previous to the exact colour in any case being noted. Where I have observed any markings left unnoticed by Mr. Darwin, I have added them myself; and, in most instances, I have given the general disposition of the colours as they appear in spirits, from the circumstance of their being often so much altered by the liquor, and liable to mislead those, who have only the opportunity of seeing them in preserved specimens. This is what Cuvier and Valenciennes have frequently done in their work ; and from them I have borrowed the practice.

In a work of this nature, it has not been thought desirable to enter into any discussion of the principles of scientific arrangement, or to effect any change in systems already received ; its main object being the description of species. For this reason, I have taken the groups almost exactly as they stand in the "Histoire des Poissons" of Cuvier and Valenciennes, or in the "Regne Animal" of the former: yet there is reason to believe that many parts of their system will be found hereafter to require some modification, especially in regard to families and genera which have for their distinctive character the presence or absence of vomerine or palatine teeth. The small value which is to be attached to such character is pointed out in some instances in the following work, and much dwelt upon.

In conclusion, it may be stated, that the whole of the species in the collection of fish brought home by Mr. Darwin, described in the following pages, have been deposited by him in the Museum of the Philosophical Society of Cambridge. They are mostly in spirit, and, generally speaking, in a good state of preservation ; some few, however, are in the state of skins only, and have been mounted.

L. JENYNS.

Swaffham Bulbeck,
Jan. 8, 1842.

SYSTEMATIC TABLE OF SPECIES,

WITH THEIR RESPECTIVE HABITATS.

ACANTHOPTERYGII.

PERCIDÆ.

Perca lævis, *Jen.*	South Patagonia.
Serranus albo-maculatus, *Jen.*	Galapagos Archipelago.
———Goreensis, *Val.?*	Cape Verde Islands.
————aspersus, *Jen.*	Ditto.
———labriformis, *Jen.*	Galapagos.
———olfax, *Jen.*	Ditto.
Plectropoma Patachonica, *Jen.*	North Patagonia.
Diacope marginata, *Cuv.*	Keeling Islands.
Arripis Georgianus	King George's Sound.
Aplodactylus punctatus, *Val.*	
Dules Auriga, *Cuv. et Val.*	Maldonado.
—— Leuciscus, *Jen.*	Tahiti.
Helotes octolineatus, *Jen.*	King George's Sound.
Aphritis undulatus, *Jen.*	Archipelago of Chiloe.
—— porosus, *Jen.*	Central Patagonia.
Pinguipes fasciatus, *Jen.*	North Patagonia.
———— Chilensis, *Val.*	Valparaiso.
Percophis Brasilianus, *Cuv.*	North Patagonia.

MULLIDÆ.

Upeneus flavo-lineatus, *Cuv.et Val.*	Keeling Islands.
—— trifasciatus, *Cuv.*	Tahiti.
—— Prayensis, *Cuv. et Val.?*	Cape Verde Islands.

TRIGLIDÆ.

Trigla Kumu, *Less. et Garn.*	New Zealand.
Prionotus punctatus, *Cuv.*	Rio de Janeiro.
—— Miles, *Jen.*	Galapagos.

COTTIDÆ.

Aspidophorus Chiloensis, *Jen.*	Chiloe.
Platycephalus inops, *Jen.*	King George's Sound.

SCORPÆNIDÆ.

Scorpæna Histrio, *Jen.*	Galapagos.
Sebastes oculata, *Val.?*	Valparaiso.
Agriopus hispidus, *Jen.*	Archipelago of Chiloe.
Apistus ——?	King George's Sound.

SCIÆNIDÆ.

Otolithus Guatucupa, *Cuv. et Val.*	Maldonado.
———— analis, *Jen.*	Coast of Peru.
Corvina adusta, *Agass.*	Maldonado.
Umbrina arenata, *Cuv. et Val.*	North Patagonia.
———— ophicephala, *Jen.*	Coquimbo.
Prionodes fasciatus, *Jen.*	Galapagos.
Pristipoma cantharinum, *Jen.*	Ditto.
Latilus jugularis, *Val.*	Valparaiso.
—— princeps, *Jen.*	Galapagos.
Heliases Crusma, *Val.*	Valparaiso.

SPARIDÆ.

Chrysophrys taurina, *Jen.*	Galapagos.

MÆNIDÆ.

Gerres Gula, *Cuv. et Val.?*	Rio de Janeiro.
———Oyena, *Cuv. et Val.?*	Keeling Islands.

CHÆTODONTIDÆ.

Chætodon setifer, *Bl.*	Keeling Islands.
Stegastes imbricatus, *Jen.*	Cape Verde Islands.

SCOMBRIDÆ.

Paropsis signata, *Jen.*	North Patagonia.
Caranx declivis, *Jen.*	King George's Sound.
—— torvus, *Jen.*	Tahiti.

b

SCOMBRIDÆ—*continued.*

Caranx Georgianus, *Cuv. et Val.* King George's Sound.
Seriola bipinnulata, *Quoy et Gaim.* Keeling Islands.
Psenes —— ? . . . . . . South Atlantic Ocean.
Stromateus maculatus, *Cuv. et Val.?* Chiloe.

TEUTHYDIDÆ.

Acanthurus triostegus, *Bl. Sehn.* Keeling Islands.
———— humeralis, *Cuv. et Val.* Tahiti.

ATHERINIDÆ.

Atherina argentinensis, *Cuv. et Val.?* Maldonado.
————microlepidota, *Jen.* . . Valparaiso.
————incisa, *Jen.* . . . . . . North Patagonia.

MUGILIDÆ.

Mugil Liza, *Cuv. et Val.?* . . . North Patagonia.
————? . . . . . . Keeling Islands.
Dajaus Diemensis, *Richards.* . . King George's Sound.

BLENNIDÆ.

Blennius palmicornis, *Cuv. et Val.* Cape Verde Islands.
Blennechis fasciatus, *Jen.* . . . Concepcion.
————ornatus, *Jen.* . . . Coquimbo.
Salarias atlanticus, *Cuv. et Val.* . Cape Verde Islands.

BLENNIDÆ—*continued.*

Salarias quadricornis, *Cuv. et Val.?* Keeling Islands
———— vomerinus, *Cuv. et Val.?* Cape Verde Islands.
Clinus crinitus, *Jen.* . . . . . Coquimbo.
Acanthoclinus fuscus, *Jen.* . . . New Zealand.
Tripterygion Capito, *Jen.* . . . New Zealand.
Iluocœtes fimbriatus, *Jen.* . . . Archipelago of Chiloe.
Phucocœtes latitans, *Jen.* . . . Falkland Islands.

GOBIDÆ.

Gobius lineatus, *Jen.* . . . . Galapagos.
Gobius ophicephalus, *Jen.* . . Archipelago of Chiloe.
Eleotris Gobioides, *Val.* . . . New Zealand.

LOPHIDÆ.

Batrachus porosissimus, *Cuv. et Val.?* Bahia Blanca.

LABRIDÆ.

Cossyphus Darwini, *Jen.* . . . Galapagos.
Cheilio ramosus, *Jen.* . . . . . Japan?
Chromis facetus, *Jen.* . . . . . Maldonado.
Scarus chlorodon, *Jen.* . . . . . Keeling Islands.
———— globiceps, *Cuv. et Val.* . . Tahiti.
———— lepidus, *Jen.* . . . . . . Tahiti.
————————? . . . . . . Keeling Islands.

MALACOPTERYGII.

SILURIDÆ.

Pimelodus gracilis, *Val.?* . . Rio de Janeiro.
———— exsudans, *Jen.* . . . Ditto.?
Callichthys paleatus, *Jen.*

CYPRINIDÆ.

Pœcilia unimaculata, *Val.* . . . Rio de Janeiro.
———— decem-maculata, *Jen.* . . Maldonado.
Lebias lineata, *Jen.* . . . . . Ditto.
———— multidentata, *Jen* . . . Monte Video.
Mesites maculatus, *Jen.* . . . . South Patagonia.
———— alpinus, *Jen.* . . . . . Tierra del Fuego.
———— attenuatus, *Jen.* . . . . New Zealand.

ESOCIDÆ.

Exocœtus exsiliens, *Bl ?* . . . Pacific Ocean.

SALMONIDÆ.

Tetragonopterus Abramis, *Jen.* . Rio Parana, S.America.
———————— rutilus, *Jen.* . . Ditto.
———————— scabripinnis, *Jen.* Rio de Janeiro.
———————— tœniatus, *Jen.* . Ditto.
———————— interruptus, *Jen.* Maldonado.

SALMONIDÆ—*continued.*

Hydrocyon Hepsetus, *Cuv.* . . Maldonado.
Aplochiton Zebra, *Jen.* . . . Falkland Islands.
———— tœniatus, *Jen.* . . . Tierra del Fuego.

CLUPEIDÆ.

Clupea Fuegensis, *Jen.* . . . . Tierra del Fuego.
———— arcuata, *Jen.* . . . . . Bahia Blanca.
———— sagax, *Jen.* . . . . . Lima.
Alosa pectinata, *Jen.* . . . . North Patagonia.
Engraulis ringens, *Jen.* . . . Coast of Peru.

PLEURONECTIDÆ.

Platessa Orbignyana, *Val.?* . . Bahia Blanca.
————————————? . . . . King George's Sound.
Hippoglossus Kingii, *Jen.* . . Valparaiso.
Rhombus ————? . . . . Bahia Blanca.
Achirus lineatus, *D'Orb.* . . . Coast of Brazil.
Plagusia ————? . . . . . . Coast of Patagonia.

CYCLOPTERIDÆ.

Gobiesox marmoratus, *Jen.* . . Archipelago of Chilôe.
———— pœcilophthalmos, *Jen.* Galapagos.

TABLE OF SPECIES.

LOPHOBRANCHII.

PLECTOGNATHI.

CYCLOSTOMI.

LIST OF PLATES.

Perca itinis. 1/2 the size.

T. Landseer sculp.

FISH.

ACANTHOPTERYGII.

FAMILY—PERCIDÆ.

PERCA LÆVIS. *Jen.*

PLATE I.

P. nigricanti-fusco undique punctata ; vertice, fronte, rostro usque ad nares, et infra-orbitalium parte posteriori, squamatis ; squamis, in capite ciliatis scabris, in corpore sublævibus.

B. 7; D. 9—1/11; A. 3/9; C. 17 ; P. 15; V. 1/5.

LONG. unc. 11 ; lin. 5.

FORM.—Much more elongated than the *common Perch*, with the back less elevated. Depth, beneath the commencement of the first dorsal, not quite equalling one-fifth of the entire length. Thickness, in the region of the pectorals, about two-thirds of the depth. Head not quite one-fourth of the entire length. Profile falling gently from the nape in nearly a straight line at an angle of about 45° : at the nape the dorsal line rises so as to interrupt its continuity with the slope of the profile, but it is nearly horizontal along the base of the dorsal fins. The jaws are nearly equal, but when the mouth is closed, the upper one appears somewhat the longer. A band of velutine teeth in each jaw, as well as on the vomer and palatines. Maxillaries when at rest nearly concealed beneath the suborbital bones : these last with their lower margin distinctly denticulated ; their surface presenting several small hollows. Eyes rather above the middle of the cheeks, and about equi-distant from the extremity of the snout and the posterior margin of the preopercle ; their diameter is one-sixth of the length of the head ; the distance from one to the other equals one diameter and a half. Nostrils double, a little in advance of the eyes ; the first orifice oval, the second round. Preopercle rectangular, with the angle rounded ;

the ascending margin finely denticulated, the teeth almost disappearing at the top ; towards the angle the teeth become stronger and point downwards ; they are also stronger and more scattered along the basal margin, inclining here a little forwards. Opercle with two flat sharp points, one a little below the upper angle, the other about the middle and terminating the gill cover. Both the subopercle and interopercle have their margins obscurely denticulated : the margin of the former is rather sinuous, and passes obliquely forwards and downwards to form a continuous curve with that of the latter. Crown, forehead, upper part of the snout as far as the connecting line of the nostrils, posterior half of the suborbitals, cheeks, and all the pieces of the gill cover, excepting the lower limb of the preopercle, covered with small scales, which are in most instances ciliated with a varying number of denticles, and feel rough to the touch : the extremity of the snout, anterior portion of the suborbitals, maxillaries, and lower jaw are naked. Above each orbit is a small semi-circular granulated plate, with the granulations disposed in striæ. The suprascapulars terminate in an obtuse projecting point. The humeral bone forms a large osseous triangular plate above the pectorals, the salient angle terminating in three small teeth. Course of the lateral line a little above one-third of the depth till it arrives beneath the second dorsal, where it bends down to half the depth. Scales on the body larger than those on the head, of an oblong form, rounded at their free edges, which are scarcely at all ciliated, and for the most part quite smooth to the touch ; their concealed portion not wider than the free, with a fan of fourteen striæ ; the rest of their surface more finely striated. The first dorsal commences a little beyond a vertical line from the termination of the humeral plate, and is almost continuous with the second, being only separated by a deep notch: the space occupied by the two dorsals together is exactly one-third of the entire length : spines strong ; the first scarcely more than one-third the length of the second, which is very little shorter than the third ; this last longest, equalling rather more than half the depth ; rest of the spines gradually decreasing to the last, which is of the same length as the first. The second dorsal commences with a slender spine, not half the length of the first soft ray, which last is simple, the others being branched ; third and fourth soft rays longest ; the succeeding ones slowly decreasing to the last, which is rather more than half the longest. Anal preceded by three spines, the first of which is very short; second much longer and very stout ; third of about the same length as the second, but much slenderer ; the first and second separated by a wide membrane from the third, which is closely united to the first soft ray ; these last longer than those of the second dorsal, but in other respects similar. The anal and second dorsal terminate in the same vertical line ; and the last ray is double in both fins. Between them and the caudal is a space equalling one-fifth of the entire length. The caudal is slightly notched. The pectorals are rather pointed, their length equalling two-thirds that of the head. Ventrals immediately beneath them, and of about the same length ; the first soft ray longest, and more than twice the length of the spine which precedes it.

COLOUR.—In spirits this fish appears yellowish brown, deepening on the back but becoming paler on the belly, and covered all over with small dusky spots, one occupying the base of each scale.

Habitat, Santa Cruz River, Patagonia.

Drawn from Nature on Stone by Waterhouse Hawkins

Serranus udomaculatus ½ Nat Size

No true perch had been obtained from South America until M. D'Orbigny discovered one in the Rio-Negro, in North Patagonia, which has been since described by Valenciennes, under the name of *P. trucha.** The present species was found dead by Mr. Darwin, high up the river of Santa Cruz, in South Patagonia. It is evidently very closely allied to the *P. trucha*, and is spotted in a similar manner; but it appears to differ in the scales not advancing on the snout beyond the nostrils, or covering more than the posterior half of the suborbitals. Those on the body are also particularly characterized by being so smooth, as hardly to communicate any sensation of roughness when the hand is passed from the tail towards the head, though the head itself is rough. This circumstance has suggested the specific name. This species further disagrees with the one above alluded to in having the caudal slightly forked, not rounded; and in having two soft rays less in the second dorsal, and one less in the anal. Valenciennes's description, however, of the *P. trucha* is very brief; on which account I have been the more minute in that of the *P. lævis*.

This perch, with *P. trucha*, would almost seem to form a subordinate division in the genus, distinguished from that embracing all the other described species, by the character of the scales covering a large portion of the head which gives it a remarkable sciænoid appearance. Both species may be known from all the North American perches, by their having the body spotted instead of banded, and by the smaller number of rays in the first dorsal. In this last character they agree with the *P. ciliata*, and *P. marginata* of Cuvier and Valenciennes.

1. SERRANUS ALBO-MACULATUS. *Jen.*

PLATE II.

S. lateribus maculis albis serie longitudinali dispositis ; dentibus velutinis ; paucis, hic et illic sparsis, fortioribus, aculeiformibus, vel sub-conicis ; preoperculo margine adscendenti convexiusculo, denticulato ; denticulis ad et infrà angulum paulò majoribus; operculo mucronibus duobus parvis, et spinâ intermediâ forti, armato ; rostro et maxillis nudis ; squamis corporis leviter ciliatis ; pinnâ caudali æquali.

B. 7 ; D. 10/13 ; A. 3/7 ; C. 17, &c.—P. 17 ; V. 1/5.

LONG. unc. 16; lin. 9.

FORM.—Of an oblong-oval form, with the greatest depth about one-fourth of the entire length. The dorsal and ventral lines are of nearly equal curvature. The profile is nearly rectilineal,

* *Hist. des Poiss.* tom. ix. p. 317. I refer to the quarto edition throughout.

falling very gradually from the commencement of the dorsal to the end of the snout, without
any elevation at the nape. The head is one-third of the entire length. The lower jaw projects
beyond the upper. The maxillary, which is broad, and cut quite square at its extremity, reaches
to beneath the middle of the orbit. The suborbital has the margin entire and nearly straight.
The upper jaw has a band of velutine teeth, broadish in front, but narrowing (the teeth at the
same time becoming smaller and finer) posteriorly; with an outer row of not much longer, but
considerably stronger, subconic teeth, placed at rather wide intervals; besides these, there are
three or four teeth on each side of the anterior portion of the jaw, equally strong as those last
mentioned, but more curved, the points reclining backwards, and set within the velutine band.
In the lower jaw, there is the same band as above, but narrower, and with the teeth more in
fine card than velutine, with stronger ones anteriorly, and along the posterior half of each side,
where there are six or eight, standing nearly in a single row, very stout and curved, though
scarcely longer than the others; outside the band, and on each side of the symphysis, there are
three or four moderately strong subconic teeth, at short distances from each other, which may
be considered as small canines. On the vomer and palatines, the teeth are velutine. The eyes
are rather large, and placed high in the cheeks; their diameter is about one-sixth the length of
the head: the distance between them equals one diameter and a quarter. The nostrils consist
of two orifices, placed one before the other, a little in advance of the eyes, roundish-oval, the
posterior one largest. The preopercle has the ascending margin not quite rectilineal, being
slightly convex, and the angle at bottom rounded; the denticulations on the former are fine,
but very perceptible; they become rather stronger and more distant at the angle, and a few of
this character are continued along the posterior half of the basal margin. The opercle is
armed with three points; the upper one is triangular, small, and not very obvious; the middle
one is a moderately strong spine, about a quarter of an inch in length; the third is a little
below this last, and resembles it in form, but is much smaller. The membrane of the opercle
terminates in a sharp angle, and is produced considerably beyond the middle spine. The line
of separation between the opercle and subopercle is not visible. The gill-opening is large,
and has seven rays. There are no scales on the snout or jaws, or between the eyes, or on the
anterior portion of the suborbital; but they are present on the cranium behind the eyes, cheeks,
(where they are numerous), and pieces of the gill-cover; the limb of the preopercle, and the
lower margin of the interopercle, however, are nearly free from them. Those on the opercle
are larger than those on the cheeks. All these scales, as well as those on the body, are finely
ciliated, communicating a slight roughness to the touch. The supra-scapular is represented by
a larger and harder scale than the rest, of a semi-elliptic form, striated on its surface, and
obsoletely denticulated on the margin. The lateral line is parallel to the back, at between one-
third and one-fourth of the depth. The pectorals are attached below the middle, of a rounded
form, the middle rays being longest, and about half the length of the head. The dorsal
commences exactly above them, and occupies a space equalling half the entire length, excluding
the caudal. The spines are sharp, and moderately strong: the first is rather more than
half the length of the second, but scarcely more than one-fifth of the length of the third,
which is longest, equalling more than half the depth of the body: from the third they decrease
very gradually to the ninth, which is of the same length as the second; the tenth is again a
little higher; this is followed by the soft rays, which are nearly even, and about one-third
higher than the last spine; the last two or three, however, are a little shorter than the others.

The anal commences in a line with the fifth soft ray of the dorsal, and ends a little before that fin: the second spine is strongest, and twice the length of the first: the soft rays are longer than those of the dorsal. There are a few minute scales between the soft rays of both dorsal and anal, to about one quarter of their height. The caudal is even, but may possibly have been worn so by use. The ventrals are directly under the pectorals, a little shorter than them, and pointed.

COLOUR.—" Varies much. Above pale blackish-green; belly white; fins, gill-covers, and part of the sides, dirty reddish orange: on the side of the back, six or seven good-sized snow-white spots, with not a very regular outline.—In some specimens, the blackish-green above becomes dark, and is separated by a straight line from the paler under parts.—Again, other specimens are coloured dirty 'reddish-orange,' and ' gallstone yellow,'* the upper parts only rather darker. But in all, the white spots are clear ; five or six in one row, and one placed above. Sometimes the fins are banded longitudinally with orange and black-green."—D.

Habitat, Galapagos Archipelago.

This species, which is undoubtedly new, was obtained by Mr. Darwin at Charles Island, in the Galapagos Archipelago. As many specimens were seen, it is probably not uncommon there. It appears to be a *Serranus*, but its canines, if they can be so called, are very small and inconspicuous. Its naked jaws require it to be placed in Cuvier's first section of that genus, though much larger than most of the species contained in it, and rather differing from them in general form. In some of its characters, it would seem to make a near approach to *Centropristes*, between which and *Serranus*, there is undoubtedly a very close affinity.

2. SERRANUS GOREENSIS. *Val.?*

Serranus Goreensis, *Cuv. et Val.* Hist. des Poiss. tom. vi. p. 384.

FORM.—The general form approaching very closely that of the *S. Gigas*. Greatest depth one-fourth of the entire length. Head rather less than one-third of the same. The diameter of the eye is one-fifth of the length of the head; and the distance from the eye to the extremity of the snout is about one diameter and a quarter. The lower jaw is covered with small scales, but not the maxillary. The nostrils consist of two round apertures, the anterior one rather larger than the posterior, and covered by a membranous flap. The teeth in the upper jaw form a velutine band, with the outer row in fine card, and two stronger and longer ones near the middle of the jaw on each side : below there is a narrow band of fine card, with stronger ones situated as above. The denticulations at the angle of the preopercle are well developed, especially two teeth which are much stronger than any on the ascending margin. The opercle has three flat spines, the middle one longest and projecting further than the others ; but the terminating angle of the membrane projects beyond this spine to a distance equalling the length of

† In this and in all other cases, Mr. Darwin has used *Werner's Nomenclature of Colours*, by Patrick Syme.

the spine itself. The dorsal has the fourth spine longest, and equalling just half the entire length of the spinous portion of the fin.   Both the spinous and soft portions have minute scales between the rays, covering rather more than the basal half of the fin ; they rise highest just at the commencement of the soft portion.   The caudal is square at the extremity, or with rather more tendency to notched than rounded ; the basal half scaly.   The anal commences in a line with the third soft ray of the dorsal, and has the basal half of the soft portion finely scaled : the second spine is strongest, but the third somewhat the longest.   The pectorals are rounded, with the seventh and eighth rays longest ; finely scaled on the upper side for one-fourth of their length from the base, but without any scales beneath.   The ventrals are a little shorter than the pectorals, with a spine of about the same length and stoutness as the third anal spine, and rather more than equalling half the length of the soft rays : they are obsoletely scaled on the upper side between the rays.

D. 11/16 ; A. 3/8 ; C. 15, &c.—P. 17 ; V. 1/5.

Length 7 inches.

Colour.—(*In spirits.*) Of a nearly uniform bister brown, stained and mottled here and there, particularly on the sides below the lateral line, with patches of a much paler tint.

Habitat, St. Jago, Cape Verde Islands.

The Serran above described, was procured by Mr. Darwin at Porto Praya. I am not sure that I am right in referring it to the *S. Goreensis* of Valenciennes, as in so extensive a genus, and one in which the species are so extremely similar, it is very difficult to identify any one in particular, without the opportunity of comparing it with a large number.   But it seems to agree with that species better than with any other I can find noticed by authors ; and the island of Goree is sufficiently near the Cape Verde Islands, to render it probable that the same species may occur in both localities.   It has the same square tail, which, according to Valenciennes, so particularly characterizes the *S. Goreensis ;* but it has one soft ray more in the dorsal.   I see no appearance of the deep violet said to border the dorsal and anal fins, but possibly it may have been effaced by the action of the spirit.

3. Serranus aspersus. *Jen.*

S. supra viridi-niger, subtus pallidior ; lateribus smaragdino pallido aspersis; pinnis anali, caudali, dorsalique postice, apicibus croceis ; dentibus velutinis, caninis in maxillâ superiore utrinque versus apicem duobus sub-fortibus ; preoperculo margine prope recto denticulato ; denticulis ad angulum paulò majoribus ; operculo mu-

cronibus tribus planis, intermedio maximo; rostro toto, et maxillâ inferiore, squa-matis.

B. 7; D. 11/15; A. 3/8; C. 17, &c.; P. 17; V. 1/5.

Long. unc. 4¼.

Form.—Back very little elevated; the greatest depth rather less than one-fourth of the entire length. Nape slightly depressed, with which exception, the dorsal line from the commencement of the dorsal fin to the crown of the head, is nearly horizontal: from between the eyes to the end of the snout, the profile is considerably convex. Head rather more than one-third of the entire length. Eyes large, their diameter about one-fourth the length of the head, high in the cheeks, and distant rather less than a diameter from the end of the snout. Lower jaw longer than the upper. The teeth above consist of a narrow velutine band, with a few, a little behind the anterior extremity, longer than the others, but slender and curving backwards; in front, and on each side of the extremity are two moderate canines: beneath there is a narrow band of velutine and fine card mixed, but no canines. The lower jaw, and the snout quite to the extremity, as well as the suborbitals, are covered with minute scales, but not the maxillary. The preopercle has the ascending margin nearly rectilineal, and finely denticulated; the angle at bottom rather sharp, and the denticles at this part, as well as immediately above it, rather more developed than the others. Opercle with three flat points; the upper and lower ones equal, the middle one larger, but not projecting so far as the membrane. Dorsal spines invested with membranous tags at their tips; of nearly equal lengths, with the exception of the first two; the third and fourth a little the longest: the soft portion of the fin higher than the spinous. Anal rounded, terminating sooner than the dorsal; the second spine a trifle longer than the third, as well as stouter. The caudal is injured, but appears to have been square, or perhaps slightly rounded. Rows of minute scales between the rays of all the vertical fins.

Colour.—" Dark greenish, black above, beneath lighter; sides marked with light emerald green: tips of the anal, caudal, and hinder part of the dorsal, saffron yellow; tips of the pectorals orpiment orange."—D. These colours have been much altered by the action of the spirit. The general ground is now dusky lead, mottled and sprinkled on the sides with dirty white. There is an appearance of four oblong black spots on the upper part of the back beneath the base of the dorsal, not noticed by Mr. Darwin. The tips of the fins have entirely lost their bright colours.

Habitat, Porto Praya, St. Jago, Cape Verde Islands.

This species was also obtained at Porto Praya, off Quail Island. It belongs to that division of the genus which Cuvier has distinguished by the name of *Mérou*, and to his section of *Mérous piquetés;* but it will not accord with any of those described in the " Histoire des Poissons." There is only one specimen of it in the collection, which is small, and probably not full-sized.

4. SERRANUS LABRIFORMIS. *Jen.*

PLATE III.

S. fusco-flavo, nigro, alboque variatus ; dorsali rubro-marginatâ ; spinis fortibus, sub-æqualibus, ad apices laciniis membranaceis investitis; dentibus aculeiformibus,* valdè retroflexis, seriebus internis majoribus; caninis, in maxillâ superiore duobus, in inferiore quatuor, mediocribus ; preoperculo margine arcuato, vix denticulato ; operculo mucrone unico plano, modico, armato; squamis infra lineam lateralem ciliatis, supra et in ventre lævibus.

B. 7; D. 11/17; A. 3.8; C. 15, &c.; P. 18; V. 1/5.

LONG. unc. 17.

FORM.—Oblong-oval, with very much the aspect of a *Labrus.* The greatest depth, which is beneath the commencement of the dorsal, is rather less than one-fourth of the entire length. The head is large, and nearly one-third of the same. The profile, from the dorsal to the end of the snout, curves gradually downwards in one continuous bend. The lower jaw projects a little beyond the upper. The teeth form a broadish band of fine card in both jaws, with the inner rows longer and more curved than the outer ; in the upper jaw, a little behind the anterior extremity, are three or four longer than the others, and curving so much backwards as almost to be laid flat; at the posterior part of this jaw on each side they pass into velutine. The canines are strong, but not very long ; in number two above and four below, not exactly in front, but a little on each side of the middle. The teeth on the vomer and palatines are velutine. The eyes are moderately large, high in the cheeks, equidistant from the upper angle of the preopercle and the end of the snout, with a diameter rather less than one-sixth the length of the head : the distance between them about equals their diameter. The margin of the suborbital is entire, but a little sinuous. The maxillary is large, and cut nearly square at its posterior extremity : it is nearly all exposed, and reaches to beneath the posterior part of the orbit. The nostrils are a little in advance of the eyes, and consist of two round openings, one before the other, the posterior one being the largest. The whole head, including the lower jaw, is covered with small scales, which become more minute towards the extremity of the snout, but are very visible even there: there are none, however, apparent on the maxillary. The preopercle has the basal angle rounded, and the ascending margin a little convex outwards, and denticulated, but the denticles are minute and not very obvious. The opercle and subopercle together (the line of separation between which is scarcely apparent) form a triangle. The former terminates posteriorly in one flat spine, moderately developed, not reaching to the extremity of the membranous angle by twice its own length. The lateral line, which is rather indistinct, is nearly parallel to the back at a little below one-fourth of the depth. The scales on the body below

---

* I have employed this term to designate the slender curved teeth, arranged in several rows, which Cuvier calls *en cardes,* or, when less numerous and rather more developed, *en crochets.* They much resemble the *prickles* found on some plants.

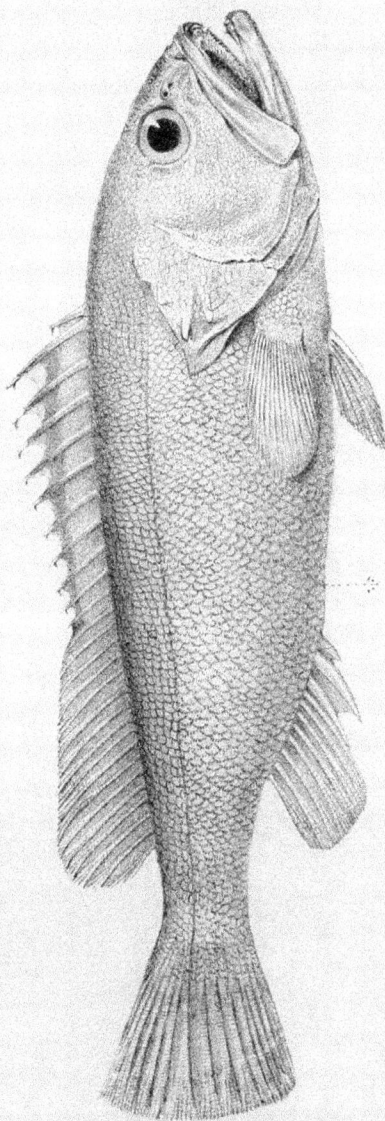

Drawn from Nature on Stone by M.Whitehead. London

Serranus labriformis ⅓ Nat Size

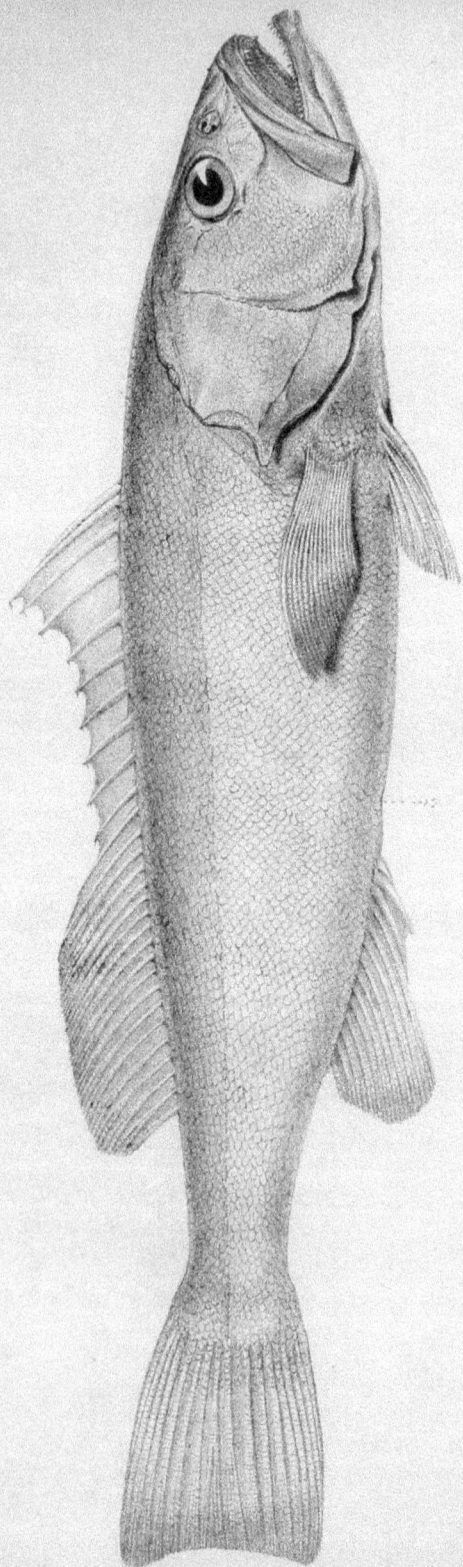

Serranus effiax P. Fez Smit

the lateral line are rough, but those above it, as well as those on the belly, nearly smooth : one taken from the middle of the side is of an oblong form, rounded at the free extremity, which is finely dotted and ciliated; its whole surface finely striated, with nine or ten deeper striæ at the base. The pectorals are attached low down, rounded at the extremity, and about half the length of the head. The dorsal commences immediately above them, and is tolerably even throughout its course, extending nearly to the caudal. The membrane is rather deeply notched between the spines, which are very stout and invested at their tips with membranaceous tags, as in the *Labridæ.* The first spine is only half the length of the second and third; the fourth is a little the longest, equalling two-fifths of the depth of the body; the fifth and succeeding ones decrease very gradually to the tenth, which is rather more than half the length of the fourth; the eleventh is a little longer, and is followed by the soft rays, the longest of which is about equal to the longest of the spinous. The anal commences about in a line with the second soft ray in the dorsal, and terminates before that fin, leaving double the distance between it and the caudal; first spine not half the length of the second and third, which are about equal, and much stouter; soft portion of the fin of a rounded form, with the middle rays nearly double the length of the second and third spines. Caudal even, or very slightly rounded, without any rows of scales between the rays. Ventrals a little shorter than the pectorals, immediately beneath them, pointed.

COLOUR.—" Mottled with brown-yellow, black and white: upper and lower edges of the caudal, edges of the dorsal and anal, ' arterial' and purplish red."—D.

Habitat, Galapagos Archipelago.

Obtained off Chatham Island in the Galapagos Archipelago. There can be little doubt of its being an undescribed species, well characterized by its *labriform* appearance, as regards the fins, rounded and nearly entire margin of the pre-opercle, and scales smooth *above,* but rough *beneath* the lateral line.

5. SERRANUS OLFAX. *Jen.*

PLATE IV.

S. fusco-variatus; spinis dorsalibus ad apices laciniis investitis; naribus orbiculatis, aperturâ unicâ magnâ, duas minores includenti ; dentibus aculeiformibus, retroflexis, seriebus paucis ; caninis, in maxillâ superiore duobus, in inferiore quatuor, cæteris vix fortioribus ; preoperculo margine adscendenti prope recto, versum angulum paulum sinuato, vix denticulato ; operculo mucronibus duobus, parvis, subæqualibus, armato ; squamis ubique lævibus.

B. 7 ; D. 11/18 ; A. 3/11 ; C. 17, &c. ; P. 17 ; V. 1/5.

LONG. unc. 23½.

FORM.—Rather elongated, with the dorsal and ventral lines equally curved, and neither departing much from a straight line. Depth, in the region of the pectorals, equalling rather more than one-

c

fifth of the entire length. Head contained three and a half times in the same. Profile sloping gradually from the commencement of the dorsal to the end of the snout in one continuous very gentle curve. The lower jaw a little the longest. The teeth are in strong card in both jaws, their points curving inwards and backwards : those above longest anteriorly, where they form about three rows ; posteriorly they become velutine, as in the last species, and consist of not more than two rows : in the lower jaw the teeth are equally large at the sides as in front, and, excepting quite at the anterior extremity, in only two rows, the inner of which is stronger than the outer. The canines are small, and scarcely stronger than the other teeth ; in number* and situation, the same as in the *S. labriformis*. The vomerine and palatine teeth are very fine velutine. Eyes rather large, and high in the cheeks, equidistant from the upper angle of the preopercle and the intermaxillary, with a diameter about one-seventh the length of the head : the distance between them equals one diameter and one-third. The margin of the suborbital is entire, and nearly straight. The maxillary, when the mouth is closed, reaches to beneath the middle of the orbit. The nostrils are a little in advance of the eyes, and consist of one large, nearly circular, aperture, enclosing two smaller ones, which are also circular and placed equally in advance. The crown, and space between the eyes, and entire cheeks, are covered with small scales ; there are also some minute ones on the lower jaw, and on the extremity of the snout before the eyes ; but they are scarcely obvious, if present, on the first suborbital, and not at all perceptible on the maxillary. The preopercle is rather more than rectangular ; the basal margin nearly straight and horizontal ; the angle rather sharp ; the ascending margin with a slight sinuosity just above the angle, afterwards straight and nearly vertical, very obsoletely denticulated throughout its course. The osseous portion of the opercle terminates posteriorly in two flat points, nearly equal, but the lower one rather the more developed, between which it is emarginate. The angle of the membrane is considerably produced beyond the lower point. The line of separation between the opercle and subopercle is tolerably obvious. Gill-opening large. All the pieces of the gill-cover are covered with scales scarcely smaller than those on the body. The scales on the body are not large, of an oblong form, with their free edges scarcely at all ciliated, not enough to feel rough to the touch ; their whole surface very finely striated, with twelve deeper striæ on the basal half, and the basal margin crenated. Lateral line not very conspicuous, parallel to the back at about one-fourth of the depth. The dorsal commences in a line with the posterior angle of the opercle, and occupies a space equalling half the entire length, caudal excluded : spines strong, and tagged at their extremities ; the second longest, equalling not quite half the depth ; third and succeeding ones gradually decreasing to the tenth, which is about half the length of the second ; the eleventh again longer ; then follow the soft rays, which are nearly even, but all higher than the last spinous. The anal commences in a line with the third soft ray in the dorsal, and terminates a little before that fin : first spine very short ; the third longest, but the second stoutest : of the soft rays the third and fourth are longest, and nearly twice the length of the third spine, being longer than the soft rays in the dorsal ; from the fourth they gradually decrease, giving this portion of the fin a rounded form. The caudal is nearly even, but the central rays are a little shorter than the outer ones. There are no rows of scales between the soft rays of the dorsal and anal, and

* There are actually only three below in this specimen, but there is little doubt of four being the normal number, one appearing to have been lost.

scarcely any trace of them between those of the caudal. The pectorals are rounded, attached low down, and about half the length of the head. Ventrals directly beneath them, shorter, and more pointed.

COLOUR.—" Mottled brown."—D. The dried skin appears nearly of a uniform brown, simply a little paler beneath. There is some indication of a whitish band along the base of the anal and soft dorsal, which may be the remains of a brighter colour. The base of the pectorals and ventrals is also paler than the extremity of those fins.

Habitat, Galapagos Archipelago.

This species was also obtained at Chatham Island, in the Galapagos, where Mr. Darwin states that it is common. In some of its characters it approaches the *S. labriformis*, but in others it is essentially different. It rather departs from most of the *Serrani* in the teeth, and in the small development of the canines. The nostrils also are rather peculiar. Perhaps it may one day be found to constitute the type of a distinct genus.

PLECTROPOMA PATACHONICA. *Jen.*

P. operculo spinis tribus, intermediâ maximâ ; preoperculo margine adscendenti denticulato ; ad angulum dente unico, et ad marginem basalem dentibus duobus, fortibus ; pinnâ dorsali spinis quartâ et quintâ longissimis ; pectoralibus radiorum apicibus e membranâ paulo exeuntibus ; caudali leviter rotundatâ.

B. 7 ; D. 13,15 vel 16 ; A. 3/8 vel 9 ; C. 17$\frac{2}{4}$; P. 17 ; V. 1 5.

LONG. unc. 15.

FORM.—Greatest depth about one-third of the entire length, excluding caudal. Head rather exceeding one-third. Profile descending obliquely in nearly a straight line from the commencement of the dorsal to the end of the snout. Eyes large, high in the cheeks ; their diameter nearly one-fourth of the length of the head. The lower jaw a little the longest : both it and the maxillary without scales. A band of velutine teeth in each jaw ; the outer row in card, with some, stronger than the others, which may be considered canines : above, the principal canines are about six in number on each side near the extremity ; below, there are three or four larger than the others similarly situated. The preopercle has the ascending margin distinctly denticulated ; on the basal margin are two strong teeth directed forwards, and a third at the angle. The opercle has three strong flattened spines ; the middle one most developed. At the lower angle of the suhopercle is a small flat moderately sharp point. Fourth and fifth dorsal spines longest ; the succeeding ones gradually diminishing to half the height of the soft portion of the fin which follows. Second anal spine very stout. Pectorals with the tips of the rays slightly projecting beyond the membrane, giving it a festooned appearance. Caudal slightly rounded.

COLOUR.—The specimen above described appears, in its present state, greyish brown, with zig-zag lines in different directions of a darker tint. A second individual is stated by Mr. Darwin to have been, when alive, " above salmon-coloured." A third is described as " above aureous-coppery, with wave-like lines of dark brown, which often collect into four or five transverse bands ; fins lead-colour ; beneath obscure ; pupil dark blue." Both these last specimens appear now, like the first, greyish-brown. The wave-like lines extend over a portion of the dorsal and anal fins.

Habitat, coast of Northern Patagonia, and the mouth of the Plata.

This species is evidently very closely allied to the *P. Brasilianum* of Cuvier and Valenciennes, and possibly may not be distinct. It differs, however, in having only two, instead of three, teeth on the basal margin of the preopercle, which character prevails in all the specimens. It has also one or two more soft rays in the anal. It likewise approaches the *P. aculeatum* of the same authors, but this last species is said to be particularly characterized by three very sharp points on the subopercle towards the lower angle, in the room of which, in the species here described, there is only one small triangular flattened point, rather sharp in two individuals, but in the third blunt, with the margin slightly crenated. The colours too appear to be different.

Mr. Darwin's collection contains three specimens, which do not materially differ from each other. The largest, measuring fifteen inches in length, was taken in forty fathoms water off the mouth of the Rio Plata. The two others, smaller, and not exceeding nine, and seven and a half inches respectively, were got on the coast of Patagonia in lat. 38° 20′: where it is stated that great numbers were obtained, many exceeding a foot in length. In these smaller specimens the canines are not so numerous or well developed as in the larger one.

" One specimen when caught, vomited up small fish and a *Pilumnus*. Was tough for eating, but good."—D.

DIACOPE MARGINATA. *Cuv.*

Diacope marginata, *Cuv. et Val.* Hist. des Poiss. tom. ii. p. 320.

FORM.—Greatest depth of the body and length of the head equal, each being not quite one-third of the entire length. Nape somewhat elevated, whence the profile falls very regularly in a slightly convex line. The jaws appear equal when open, but when closed the upper one is a little the longest. Teeth velutine, with four well-marked canines in the upper jaw, two on each side of the anterior extremity, the outer one of which is longer than the inner. Opercle with two flat blunt points. Denticulations of the preopercle, particularly those at the angle below the notch, moderately well developed. Tubercle of the interopercle prominent. There are scales on the cheeks and pieces of the gill-cover, but none on the crown, snout, jaws, or suborbitals. The scales on the body are moderately large. There are rows of small scales between the rays of the vertical fins, but they are more developed between the soft rays than between the spinous. The dorsal has the first spine half the length of the second, which itself is a little shorter than the third ; fourth, fifth, and sixth equal and longest ; there is very little difference in the lengths of the remaining rays, nor is there much between the spinous and soft portions of the fin, which, taken as a whole, appears nearly even throughout. Anal short, commencing in a line with the fourth soft ray of the dorsal, and terminating at the same distance from the caudal as that fin : second and third spines very stout. Pectorals narrow and pointed, a little shorter than the head. Ventrals a little shorter than the pectorals.

B. 7 ; D. 10/14, the last double ; A. 3/8, the last double ; C. 17, and some short ones ; P. 16 ; V. 1/5.

Length 6 inches.

COLOUR.—" Upper part pale lead colour : pectorals yellow ; ventrals and anal orange : sides very pale yellow."—D. In spirits, the colour appears almost uniform greyish-white. The dorsal and anal fins have an edging of black, which is not noticed by Mr. Darwin, and which is characteristic of the species. The caudal is entirely dusky. There are no traces of spots on any part of the body.

Habitat, Keeling Island, Indian Ocean.

All the known species of *Diacope* are stated by Cuvier and Valenciennes, as coming from the Indian seas. The *D. marginata* was first brought from thence by Commerson. It was afterwards received by the authors above mentioned from Pondicherry. The expedition under Captain Duperrey, met with it at the Island of Oualan. Mr. Darwin's specimen was obtained at the Keeling or Cocos Islands : I believe it to be referrible to this species, as it possesses the characteristic black edging on the dorsal and anal fins ; but as the description in the "Histoire des Poissons" is very brief, containing a mere notice of the colours, I have thought it advisable to annex that of the present individual.

GENUS—ARRIPIS. *Jen.*

Membrana branchialis septem-radiata ; aperturá amplá. Pinna dorsalis unica; spinis gracilibus. Dentes aculeiformes, tenues; serie maxillari externá cæteris fortiori. Operculum mucronibus duobus parvis posticè urmatum. Preoperculum marginibus basali et adscendenti denticulatis; spinis nullis. Ossa infraorbitalia leviter denticulata. Os maxillare squamosum. Squamæ corporis levissimè ciliatis, posticè striis levissimis, transversis, parallelis, flabelli locum occupantibus.

I propose to establish this new genus for the reception of the *Centropristes Georgianus* of Valenciennes, which appears to offer sufficient peculiarities to warrant such a step. Its herring-like form, denticulated suborbital, scaly maxillary, small pectorals, backward position of the ventrals, and deeply forked caudal, sufficiently distinguish it from *Centropristes*, with which it hardly agrees in any of its characters, beyond that of wanting canines, and having the preopercle denticulated, and the opercle armed with small sharp points. Its teeth, however, are not exactly velutine, as in the typical species of that genus, but rather in fine card, with the outer row in both jaws stronger than the others. But, perhaps, one of the most marked peculiarities in this proposed genus resides in the scales, which have, instead of the usual fan of diverging striæ on their basal portions, a triangular space filled up by a number of extremely fine, closely-approximating striæ, parallel to each other, and also parallel to the basal margin, which is cut quite square and entire.*

Although this genus is thus separated from *Centropristes*, there is no doubt of its having a near affinity with it; and also with *Grystes*, from which last, however, it is at once known by its denticulated preopercle. It is still more closely allied to *Apsilus*, which it very much resembles in its general form, as well as in some of its particular characters. Amongst other points of resemblance with this last genus, may be noticed the similarity of the teeth; the very large gill-opening; the small and inconspicuous points on the opercle; the weak spines of the dorsal and anal, both which fins also terminate in a point behind; the small pectorals, and the deeply forked caudal.

It is probable that the *Centropristes truttaceus* of Cuvier and Valenciennes also belongs to this new genus, which, as well as the *C. Georgianus*, comes from New Holland, and which those authors seem, not without much hesitation, to have placed provisionally in the group in which it now stands. It is not stated, however, whether the peculiar character of the scales in the *C. Georgianus*, above pointed out, exists also in this species.

ARRIPIS GEORGIANUS.

Centropristes Georgianus, *Cuv. et Val.* Hist. des Poiss. tom. vii. p. 338.

FORM.—As M. Valenciennes has given an accurate and detailed description of this fish, and as I
have already stated above some of its leading characters, it is not necessary to say much
further on this head in reference to the specimen in Mr. Darwin's collection. I need only
point out wherein it differs from the description in the "Histoire des Poissons," the greater
part of which applies exactly. M. Valenciennes states that the ventral profile is more curved
than the dorsal, but there is not much difference in their respective degrees of curvature in this

* The absence of the usual fan has suggested the name of *Arripis*, from α priv. et ριπις, flabellum.

specimen. The thickness of the body, which he fixes at one-third of the depth, is here nearly half the depth. The following characters may be also given, in addition to his. Above each orbit are two short crests or ridges which meet at an angle anteriorly, and the interocular space between these pairs of ridges is rather depressed ; beyond, or immediately above the upper lip, the snout is a little protuberant. The band of teeth in each jaw is narrow, with the outer row longer than the others ; and at the sides of the jaw, this outer row is all that is obvious· The intermaxillary is slightly protractile. The eye is hardly removed so much as one diameter from the end of the snout. The limb of the preopercle is striated ; the angle at bottom rounded, and much dilated, so that the ascending margin falls in advance of a vertical. The contour of the membrane of the opercle is rounded. The cheeks, and all the pieces of the gill-cover, with the exception of the broad limb of the preopercle, are scaly : there are also a few scales on the maxillary, but none on the crown between the eyes, or on the snout, or lower jaw. The dorsal and anal terminate nearly in the same vertical line, but the latter reaches a trifle the farthest. Both fins are invested at their base with a scaly membrane, the scales of which are of a long lanceolate form. The length of the caudal equals the depth of the body. That of the pectorals equals half the depth : these fins are attached a little behind the opercle, and a little below the middle. The point of attachment of the ventrals is in a vertical line which passes through the middle of the pectorals, and coincides with the commencement of the dorsal. They are longer than the pectorals ; and in their axillæ is a lanceolate membranaceous scale half their own length. There is a similar, but shorter scale in the axilla of the pectorals also.

B. 7 ; D. 9/16 ; A. 3/10 ; C. 17$\frac{6}{8}$; P. 15 ; V. 1/5

Length 9 inches 10 lines.

COLOUR.—Not noticed in the recent state. In spirits, the whole fish appears of a nearly uniform dull metallic yellowish-white, tinged with olive on the back and upper part of the sides.

Mr. Darwin obtained this species in King George's Sound, in New Holland, the same place in which it was discovered by MM. Quoy and Gaimard.

APLODACTYLUS PUNCTATUS. *Val.*

Aplodactylus punctatus, *Cuv. et Val.* Hist. des Poiss. tom. viii. p. 352. pl. 242.

This very remarkable fish was first sent from Valparaiso, by M. D'Orbigny, where it was also observed by M. Gay. Mr. Darwin's collection contains a specimen, which has unfortunately lost the number attached to it ; but as he made a collection on that coast, it was probably obtained in the same locality. The description given of it in the " Histoire des Poissons," is so detailed as well as accurate, and the figure so exact, that it is quite unnecessary to annex that of the present individual. I may merely observe that the number of simple rays at the bottom of the pectorals, which appears to be a character of some importance, and which has led to the generic name of *Aplodactylus*, amounts in this specimen to six, being two more than was observed by M. Valenciennes in his, though the total number of rays in this fin is the same. I may also allude to the circumstance of the dorsal being invested at the base with a thickened membrane on each side, closely covered with small scales, which extends over nearly its whole length, but is most conspicuous along the spinous portion. This character is not mentioned by M. Valenciennes. Neither does he mention the rows of minute scales, which occur between the rays of all the fins, except the ventrals.

Mr. Darwin's specimen of this fish is eleven inches in length. The following is the fin-ray formula:

B. 6; D. 15—1/21; A. 3/8; C. 17. &c.; P. 9—vi; V. 1/5.

1. DULES AURIGA. *Cuv. et Val.*

Dules Auriga. *Cur. et Val.* Hist. des Poiss. tom. iii. p. 83. pl. 51.

FORM.—This species is remarkable for the prolongation of the third dorsal spine, which, in the present specimen, is not quite equal to half the entire length of the head and body; a small portion, however, appears to have been broken off. The greatest depth is contained three and a half times in the entire length. The head, measured to the extremity of the opercular membrane, exactly equals the depth. The line of the profile is not quite straight, there being a slight depression at the nape, above which is a convexity in immediate advance of the dorsal fin. The lower jaw is a very little the longest. The eyes are large; and the distance between them barely equals their diameter. The other characters are exactly as stated in the " Histoire des Poissons."

B. 6; D. 10/13; A. 3/7; C. 17; P. 17; V. 1/5.

Length 5 inches 3 lines.

COLOUR.—The recent colours are given by Mr. Darwin in his notes as follows : " Sides with numerous waving longitudinal lines of brownish red; the intermediate spaces greenish-silvery, so figured as to look mottled. Head marked with lines of dull red and green. Ventral and anal fins dark greenish blue."—He does not notice the vertical bands alluded to by Cuvier and Valenciennes, which are sufficiently obvious, and which accord with the figure and description of the authors just mentioned.

Habitat, Maldonado Bay, Rio Plata.

FISH.

2. DULES LEUCISCUS. *Jen.*

Dules malo, *Val. ?* Hist. des Poiss. tom. vii. p. 360.

D. pinnis caudali, anali, dorsalique molli, nigro-marginatis; dorsali profundè emarginatá, spiná ultimá radiis articulatis breviore; operculo mucronibus duobus, inferiore maximo, armato; preoperculo margine adscendenti levissimè denticulato, basali denticulis fortioribus.

B. 6; D. 10/11; A. 3/12; C. 16, &c.; P. 13; V. 1/5.

LONG. unc. 4. lin. 5.

FORM.—General form resembling that of a small *Dace*. Greatest depth about one-third of the entire length, caudal excluded. Length of the head rather less. Dorsal line falling with the profile in one continuous gentle curve. Eyes large; their diameter contained two and a half times in the length of the head : the distance between them less than one diameter. Suborbitals finely but very conspicuously denticulated. Jaws nearly equal; the lower one a little the longest. In each, a band of velutine teeth, with the outer row rather longer than the others. Opercle with two points, the lower one most developed. Preopercle with the limb striated : the ascending margin with the denticulations so fine as to be hardly sensible to the naked eye; those on the basal margin larger and more obvious. Scales of a moderate size; about forty-three in a longitudinal row; their free portions finely striated. Cheeks and opercle scaly; crown naked, with a shallow groove above each eye. Lateral line at first slightly descending, but afterwards straight; its course, until past the dorsal and anal fins, a little above the middle of the depth. Dorsal deeply notched : the anterior portion consisting of nine spines; the first very short, and scarcely more than half the length of the second; third and fourth longer, increasing gradually; fifth and sixth equal and longest, equalling half the depth of the body; seventh, eighth, and ninth, shorter, and gradually decreasing; the tenth spine, with which the second portion of the fin commences, is of the same length as the fifth, but not quite so long as the soft rays which follow; these soft rays, however, gradually become shorter, the last two not more than equalling the second spinous. The whole space occupied by the dorsal is more than one-third of the entire length. Anal commencing in a line with the ninth dorsal spine; its own three spines gradually increasing in length, but the second the strongest; soft portion of this fin longer than the corresponding portion of the dorsal, and terminating a little nearer the caudal. Vent in a line with the seventh dorsal spine. Pectorals small, reaching to the vent. Ventrals attached a little further back, and reaching a very little beyond it. Caudal forked.

COLOUR.—*(In spirits.)* Silvery, tinged on the back, and above the lateral line, with bluish grey, and somewhat mottled in places with darker spots. Fins yellowish-grey, tinged with dusky. The caudal, anal, and soft portion of the dorsal, are a little mottled with dusky, besides having a black edging; there is also a conspicuous black spot at the anterior angle of this last fin.

D

A second specimen is only three inches and a quarter in length ; but differs in no respect from the above, except in having one soft ray less in the anal fin.

Habitat, River Matavai, Tahiti.

Several of the species in this genus are extremely similar as well in form as in colours. Possibly that which I have here characterized as new may not be distinct from the *D. malo* of Valenciennes, which comes from the same country ; but the description in the " Histoire des Poissons" is so brief, that it is hardly possible to determine this point with certainty. It has, however, two, and one specimen three, soft rays less in the anal fin. It is also closely allied to the *D. marginatus*, from which it hardly differs, excepting in having the denticulations of the preopercle rather stronger, and the tenth dorsal spine shorter in relation to the soft rays which follow. The *D. marginatus*, however, comes from Java. The species here described was found by Mr. Darwin in Tahiti, in the river of Matavai.

HELOTES OCTOLINEATUS. *Jen.*

H. corpore lineis longitudinalibus nigricantibus octo ; pinnis dorsali, anali, caudalique, maculis fuscis ; vertice striis elevatis duobus subparallelis ; preoperculo distinctè denticulato, et ad marginem limbi internum subcristato ; operculo mucronibus duobus, superiore minimo ; squamis ubique lævissimis.

B. 6 ; D. 12 9 ; A. 3 7 ; C. 17, &c. ; P. 15 ; V. 1 5.

Long. unc. 9. lin. 9.

Form.—Body oblong. Greatest depth exactly four and a half times in the entire length. Length of the head rather less than the depth of the body. Snout short and obtuse. Jaws exactly equal : each with a broadish band of velutine teeth, which are all, apparently even the most minute, three-pointed, although this character is not very obvious except in the outer row, which are longer than the others. No vomerine teeth appear externally, but they may be felt through the skin of the palate, and on dissecting this off, there is brought to view a small hard disk rough with minute asperities. Mouth very little cleft, the commissure not extending more than half way between the end of the snout and the anterior margin of the orbit. Eyes rather large ; their diameter one-fourth the length of the head. Maxillary, when the mouth is closed, concealed in part beneath the suborbital, the lower margin of which is somewhat sinuous and obscurely denticulated, the denticulations being concealed by the membrane and more easily felt than seen. The denticulations on the preopercle very manifest. The principal spine on the opercle slender and very sharp, not exactly straight, but slightly curved, the convexity of the bend being downwards ; above is a second spine, but very small and easily overlooked.

The crown of the head has two nearly parallel elevated lines, which take their origin between the nostrils, and terminate at the occiput, but do not meet as in the *H. sexlineatus*; * a third line commences there exactly between them, and runs singly in a backward direction down the middle of the nape; this last is scarcely more than half the length of the two former. The cheeks and pieces of the gill-cover are scaly; but not the cranium, snout, jaws, or limb of the preopercle, which last is margined internally by a slightly elevated ridge. The scales on the body are thin and small, and without any trace of denticulations on their free edges, even under a magnifier, and the body of the fish is quite smooth to the touch rubbed either way. Lateral line as in *H. sexlineatus*. Dorsal also nearly similar, but more deeply notched, the membrane beyond the eleventh spine falling nearly to the base of the twelfth, which precedes the soft portion : sixth spine longest, equalling very nearly, but not quite, half the depth; the eleventh equals the second; the twelfth is about one-third longer than the eleventh, but is itself scarcely half the first soft ray. The anal has three soft rays less than the *H. sexlineatus*, and there are apparently but two spines, the first being (at least in this specimen, where, however, there may have been a portion broken off) quite short and rudimentary; the second and third spines are both slender, the former being rather more than half the length of the latter, and this last rather more than half the first soft ray. All the fins take their origin as in *H. sexlineatus*. The pectorals are about two-thirds the length of the head. The ventrals, which are very near together, are longer than the pectorals, but do not equal the head : they have no elongated scale between them, or in their axillæ.

COLOUR.—For the most part similar to that of the *H. sexlineatus*; but the longitudinal dark lines are more numerous, amounting to eight, with faint traces of a ninth: the additional ones are on the upper half of the sides, or above the lateral line, there being four (instead of two) above that one which passes through the eyes; the sixth extends the whole length of the fish from the end of the maxillary to the base of the caudal; the seventh passes immediately below the pectoral, and terminates in advance of it, without quite reaching to the edge of the gill-cover; the eighth is exactly equi-distant from the pectoral and ventral; this last is a very narrow pale line, but the others, with the exception of the first two, are broader and well marked. The soft portion of the dorsal, as well as the anal and caudal are spotted; the spots on this last unite to form transverse fasciæ; those on the anal are not very well-defined. The pectorals and ventrals are without spots, and pale.

Habitat, S. W. coast of Australia.

This species was procured in King George's Sound, New Holland. It closely approaches the *H. sexlineatus* of Cuvier and Valenciennes, the only species of the genus hitherto described, and obtained in the same seas by MM. Quoy and Gaimard. I have little hesitation, however, in pronouncing it to be distinct. Independently of the additional longitudinal lines on the body, and the spots on the fins, which, it is expressly stated by the above authors, are not present in the *H. sexlineatus*, it is distinguished by the striæ on the crown not meeting behind, the dorsal being rather more deeply notched, and the fin-ray formula different.

* Or at least as represented in the figure in the Histoire des Poissons, tom. iii. pl. 56.

There is one soft ray less in the dorsal, and three less in the anal; also the first spine in this last fin, if it be not broken off in this specimen, is quite rudimentary. The scales present no trace of cilia on their free edges.

1. PINGUIPES FASCIATUS. *Jen.*

PLATE V.

P corpore fasciis transversis duodecim castaneo-fuscis, alternis latioribus; dentibus palatinis paucis, conicis, subaggregatis, vix seriem formantibus; pharyngalibus aculeiformibus; membrana branchiali mediocriter emarginata; operculo spina unica forti, altera superiore obsoleta; pinnis ventralibus accuratè thoracicis.

B. 6; D. 7/27; A. 1/24; C. 15, &c.; P. 18; V. 1.5.

LONG. unc. 12. lin. 9.

FORM.—Body thickest, as well as deepest, in the region of the pectorals, compressed behind, and becoming more so as it approaches the tail; depth also gradually diminishing from that point. The greatest depth is rather less than five and a half times in the entire length : head contained four times and a quarter in the same. The thickness at the pectorals is at least three-fourths of the depth : and the thickness of the head is quite equal to it. Dorsal line nearly straight from behind the eyes, in front of which the profile descends obliquely. Eyes high, nearly reaching to the line of the profile; a little behind the middle point of the length of the head; their diameter rather less than one-fifth of this last ; the distance between them one diameter and a half. The commissure of the lips does not reach to the eyes by a space equalling half the diameter of the eye. Jaws equal. Lips very thick and fleshy, and partially reflexed, like those of a *Labrus*. Teeth very similar to those of that genus. In the upper jaw, an outer row (extending all round) of strong, sharp, slightly curved teeth, regularly set, and nearly even, but with the anterior ones a little the longest ; in all about forty, twenty on each side ; behind these a velutine band, broadest in front, but also extending the whole way round. In the lower jaw, a row of curved strong teeth, similar to those above, but extending only half way along the sides of the jaws (about nine on each side), and followed by about seven or eight short blunt conical ones ; a broad velutine band behind the longer curved teeth, but not behind the others. On the front of the vomer are four or five large blunt conical teeth, mixed with smaller ones of the same form: there is also a small group of these little conical teeth at the commencement of each palatine, but they are not carried on further in a single row.* Tongue small and inconspicuous, fastened down except just at the tip, smooth. Pharyngeal teeth in strong card; but no conical ones behind that are visible. Branchial membrane united to its fellow, and free all round at the margin, with a moderately deep notch underneath. Preopercle rounded at the angle; the ascending margin oblique. Opercle with a strong sharp spine at its upper angle, but not ex-

* As described by Cuvier and Valenciennes to be the case in the *P. Brasilianus.*

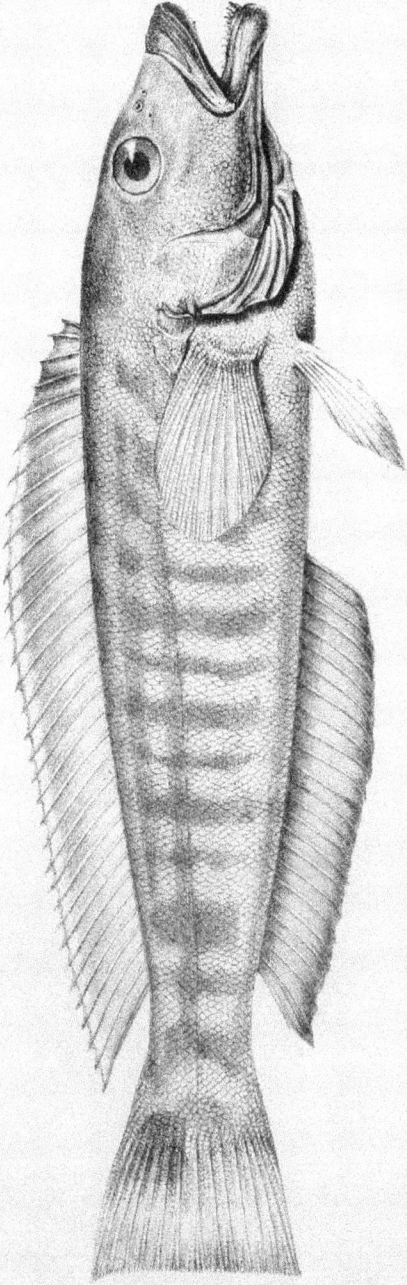

Percophis brasilianus. ¾ the size

tending beyond the membrane; a second rudimentary one above it obtusely rounded. Small scales on the cheeks, preopercle, and opercle, but not on the snout, or between the eyes, or on the suborbital, or jaws, or branchial membrane, or interopercle. The scales on the body are rather small, finely ciliated on their edges, thin, and of an oblong form, cut square at the basal margin, with a fan of twelve or fifteen striæ. Lateral line not very strongly marked, taking nearly a straight course from the upper part of the scapular to the caudal. No particular lines, markings, or pores, about the head, jaws, or between the eyes. Pectorals rounded; two-thirds the length of the head. Ventrals exactly beneath them, a very little shorter, thick and fleshy, so that the rays can hardly be distinguished. Dorsal and anal similar to those of the *P. Brasilianus;* the former has the spinous rays at first low, but the rest of the fin is of one uniform height, equalling a little less than half the depth : the latter commences under the sixth soft ray of the dorsal, and terminates in the same line. Caudal square, with rows of small scales between the rays for half their length : also a few minute scales at the base of the pectoral rays, but none on the other fins.

Colour.—" Above pale ' chestnut brown,' so arranged as to form transverse bands on the sides ; sides, head, fins, with a black tinge ; beneath irregularly white : under lip pink : eyes with pupil black, and iris yellow."—D. *In spirits;* the back and upper half of the sides are brown, the lower half of the sides and belly pale, with twelve transverse dark fasciæ, the alternate ones broader than the others. The dorsal and anal appear to have been bluish, the tint increasing in intensity from the base upwards ; but the former is edged above with a narrow white line just beneath the tips of the rays, which extends the whole length of the soft portion of the fin. The inside of the ventrals appears also to have been bluish ; but the pectorals are pale, or yellowish. Caudal brown like the back.

Habitat, coast of Northern Patagonia.

From the east coast of Patagonia, in Lat. 37° 26'. There can be no doubt of its belonging to the genus *Pinguipes,* with which it agrees in its very strong re-semblance to the *Labridæ,* as regards the head, lips, and teeth, and in its fleshy ventrals ; but there are very few teeth on the palatines, seeming to show that there is not much ground for separating this genus from *Percis.* In many of its characters, it resembles the *P. Brasilianus* of Cuvier, but it is decidedly distinct in others. It differs slightly in its proportions ; in the palatine and pharyngeal teeth ; in the position of the ventrals, which are not at all jugular, but imme-diately beneath the pectorals ; in the branchial membrane being more notched ; and in having two soft rays less in the anal. The colours also are different.

This fish is so like a *Labrus,* that at first sight it might easily deceive a student. Nevertheless its vomerine teeth, spines on the opercle, and ciliated scales, point out its right family. At the same time no system can be considered natural, which does not admit *Pinguipes* as one of the connecting links between the *Per-cidæ* and *Labridæ.*

2. PINGUIPES CHILENSIS. *Val.*

Pinguipes Chilensis, *Cuv. et Val.* Hist. des Poiss. tom. ix. p. 338.

FORM.—More slender and elongated than the last species. Depth nearly six and a half times in the entire length. Head four times and a quarter in the same. Eyes high, a little before the middle, or with the distance in front to the end of the snout not equalling that behind measured to the posterior part of the opercle; their diameter nearly six times in the length of the head; the interval between them nearly two diameters. When the mouth is closed, a vertical from the posterior part of the maxillary forms a tangent to the anterior part of the orbit. Lips not so thick and fleshy as in the *P. fasciatus*; but the teeth almost exactly similar. Tongue much larger, occupying nearly the entire platform of the mouth. Branchial membrane much more deeply notched, the notch reaching as far as the anterior extremity of the interopercle. Preopercle with the ascending margin nearly vertical. Opercle with two small flat spines, the lower one rather more developed then the upper. Scales and lateral line as in the *P. fasciatus*. Pectorals similar. Ventrals attached entirely in front of the pectorals, though not much in advance; fleshy, but perhaps rather less so than in the *P. fasciatus*: in neither species do they pass beyond the pectorals, or indeed reach quite so far. The other fins exactly similar. The dorsal, however, has one spine less, and one soft ray more. The anal, also, has one soft ray more.

B. 6; D. 6 28; A. 1 25; C. 17, &c.; P. 19; V. 1 5.

Length 11 inches.

COLOUR.—(*In spirits.*) Back and sides deep brown, with the exception of two rows of pale spots along the sides, very faint and ill-defined. Underneath altogether paler. The dorsal and anal appear to have been bluish, with the basal portion of each fin pale, but without any edging of white above. Inside of the ventrals blue; pectorals the same, but paler. The caudal shows some trace of a dark round spot on the base of the upper lobe. Mr. Darwin's notes, with respect to the colour in the living fish, only state " fins dark."

Habitat, Valparaiso, Chile.

This species, which was procured by Mr. Darwin at Valparaiso, is probably the same as the *P. Chilensis* of Valenciennes, obtained by M. Gay on the same coast. But the description in the "Histoire des Poissons" is brief, and notices very little besides the colours, which accord tolerably well. Mention, however, is made of a second spine in the anal fin, which certainly does not exist in the above specimen, though a very careful examination was made in search of it. There is also one soft ray more in this fin, as well as in the dorsal, in the fin-ray formula in that work.

This species is very distinct from the *P. fasciatus* last described, and does not show so strong a resemblance to the *Labridæ*; but it approaches very closely the *P. Brasilianus*.

Percophis Brasilianus. *Cuv.*

Percophis Brasilianus, *Cuv. et Val.* Hist. des Poiss. tom. iii. p. 209. pl. 64.
———— Brasiliensis, *Freycinet*, (Voyage) Zoologie, p. 351, pl. 53. fig. 1.

Form.—Depth and breadth in the region of the pectorals about equal, each being contained ten and a half, or nearly eleven times in the entire length. Head not quite four and a half times in the same. In the upper jaw, three strong, curved, sharp-pointed canine teeth on each side; besides a velutine band extending the whole way, with the outer row longer and more developed than the others : in the lower jaw a velutine band, with long sharp canines, similar to those above, arising amongst them at nearly regular intervals, to the number of ten or eleven on each side; none exactly in front, and not all of the same size, but passing here and there into card. Membranous margin of the preopercle very finely, almost obsoletely denticulated. Branchial membrane with seven rays, the seventh being not much smaller than the sixth.* The whole head covered with scales, including the lower jaw, and the upper half of the maxillary. Lateral line nearly straight, a little above the middle. First dorsal commencing at about one-third of the entire length, excluding caudal; of a triangular form, with its length a little exceeding its height; second spine longest, about equalling the depth of the body. Distance between the two dorsals equalling half the length of the first. Second dorsal with the first ray longest, equalling the longest of the spines in the first dorsal; second and succeeding rays slightly decreasing to the sixth, beyond which they are nearly even, with the exception of the last three or four, which are shorter; all these rays very much branched, with the intervening membrane deeply notched. Anal commencing a little before the end of the first dorsal, and terminating nearly in a line with, but in strictness a very little beyond, the second dorsal; rays and membrane much as in that fin, to which it answers in general height. Distance between the second dorsal and caudal, only one-twenty-eighth of the entire length. Caudal appears obliquely square, the upper rays being slightly longer than the lower, but perhaps worn so. Pectorals one-eighth of the entire length. Ventrals about three-fourths of their length, attached in front of them, as described by Cuvier. In the axillæ of the pectorals is a falcated membranaceous appendage covered with scales (not noticed by Cuvier), a very little less than one-fourth the length of the fins themselves.

B. 7; D. 10—32; A. 41; C. 15. &c.; P. 18; V. 1/5.

Length 21 inches.

Colour.—"Above pale, regularly and symmetrically marked with brownish red, the tip of each scale being so coloured. Beneath silvery white. Sides with a faint coppery tinge. Ventral fins yellowish. Pupil of the eye intense black."—D.

* Cuvier in his description, says, of the seventh ray, "fort petit," but it is very obvious in this specimen, and scarcely smaller than the sixth, as above stated.

Second Specimen.—Breadth or thickness at the pectorals about ten and a half times in the entire length. Depth at that point less than the breadth. Canine teeth in the lower jaw smaller than those above, and not set at such regular intervals as in the first specimen.[*] Scarcely any appearance of denticulations on the membranous border of the preopercle. Distance between the two dorsals a little less than the length of the first. Pectorals contained eight and a half times in the entire length. Fin-ray formula as follows :—

D. 9—32; A. 42 ; C. 15, &c. ; P. 17 ; V. 1.5.

Length 14 inches.

In all other respects exactly similar to the specimen first described.

Habitat, coast of Northern Patagonia, and Maldonado.

Mr. Darwin's collection contains two specimens of this fish, which was first discovered by MM. Quoy and Gaimard at Rio Janeiro. The larger one was caught by hook and line in fourteen fathoms water on the coast of Patagonia, in lat. 38° 20'. The second was taken at Maldonado, where he states it to be common. They differ in several respects from the description and figure in the "Histoire des Poissons," of Cuvier and Valenciennes ; but as they also differ a little from each other, the species is perhaps subject to variation. Amongst other points, I may mention the scales on the jaws, which are expressly stated by Cuvier to be *without* scales ; and also the emargination of the membrane between the rays of the second dorsal and anal, which is not represented in his figure, nor alluded to in his description, though very striking. This last character appears, however, in the figure given in the Zoological Atlas of Freycinet's voyage, which is on the whole a more correct representation. "When cooked, was good eating."—D.

FAMILY.—MULLIDÆ.

1. UPENEUS FLAVOLINEATUS. *Cuv. et Val.*

Upeneus flavolineatus, *Cuv. et Val.* Hist. des Poiss. tom. iii. p. 336.

FORM.—Considerably elongated. Greatest depth contained five times in the entire length, caudal excluded. Head three and a half times in the same. Dorsal line nearly straight. Profile very convex. Crown between the eyes broad and somewhat depressed, forming a slight hollow.

---

[*] Probably these teeth are liable to be lost or broken off, so as seldom to occur in exactly the same number and mode of arrangement in two individuals.

Eyes large; their diameter more than one-fourth that of the head. Suborbitals marked on their surface near the lower margin with six or eight diverging salient lines, each terminating at bottom in a mucous pore. Teeth forming a narrow velutine band, hardly visible to the naked eye, but sensible to the touch: none on the vomer or palatines. Opercle with one short flat spine projecting beyond the posterior margin rather more than half a line. Barbules reaching to a little beyond the angle of the preopercle. Mucous tubes of the lateral line with five or six branches; the branches not always simple,* but consisting sometimes of two or three main ones which are subdivided. First dorsal of a triangular form, with the spines rather slender; the first two equalling more than three-fourths of the depth of the body. Space between the dorsals about equalling the length of the first. Second dorsal with the first ray (or spine) scarcely more than half the length of the second, which is longest; the third and succeeding rays gradually decreasing to the last, which is shortest. Length of the second dorsal just equalling its greatest height. Anal answering to this last fin. Caudal deeply forked; the central rays not being one-fourth the length of the outermost ones. Ventrals and pectorals exactly of the same length; both reaching to a vertical line from the extremity of the first dorsal. Vent in a line with the commencement of the second dorsal.

D. 7—1/8; A. 1/6; C. 15, &c.; P. 16; V. 1/5.

Length 6 inches 9 lines.

COLOUR.—" Dull silvery, with a yellow stripe on the side."—D.

There can be but little doubt of this species being the *U. flavolineatus*, which appears to have a wide range over the Indian Ocean, and also to occur in the South Pacific. Mr. Darwin's specimen was taken at the Keeling Islands.

2. UPENEUS TRIFASCIATUS. *Cuv. et Val.*

Upeneus trifasciatus, *Cuv. et Val.* Hist. des Poiss. tom. iii. p. 344.

FORM.—General form resembling that of the *Mullus Surmuletus*, but the snout more elongated. Greatest depth contained about four times and a half in the entire length. Head exactly one-fourth of the same. Eyes small, distant three diameters from the end of the snout. Suborbitals with a moderate number of pores on their disk, but without any salient lines. Posterior extremity of the maxillary broad. A single row of conical teeth in each jaw, very uniform in size, not very large or very close; about twenty-two above and twenty below. Spine of the opercle about a line in length. Barbules reaching to, or a little beyond, the posterior margin of the opercle. Lateral line not much ramified. Height of the first dorsal equalling more than half the depth. Space between the two dorsals equalling one-third the length of the second dorsal.

* As stated by Cuvier and Valenciennes.

E

Both this last fin and the anal terminating in a considerable point behind. Ventrals large, reaching very nearly to the anal.

D. 8—9 ; A. 7 ; C. 15, &c. ; P. 16 ; V. 1/5.

Length 7 inches 9 lines.

COLOUR.—(*In spirits.*) Dark brownish yellow, with faint indications of three dusky patches or abbreviated transverse fasciæ, one beneath each dorsal, and the third on each side of the upper part of the tail. Second dorsal and anal crossed by several whitish longitudinal lines; the posterior point of each fin nearly black.

This species was obtained by Mr. Darwin at Tahiti. It is probably the *U. trifasciatus* of Cuvier and Valenciennes, who received their specimens from the Carolinas and Sandwich Islands. But it does not so well accord with the *Mulle multibande* of Quoy and Gaimard, which is supposed by the authors of the " Histoire des Poissons," to be the same as their species. If the figure in the Zoology of " Freycinet's Voyage" be correct, the *Mulle multibande* has the nostrils much smaller, and the spines of the first dorsal much stronger ; the ventrals also are relatively much shorter, so as to reach very little more than half way to the anal. Future observation must determine whether the two fish are distinct or not.

3. UPENEUS PRAYENSIS. *Cuv. et Val.?*

Upeneus Prayensis, *Cuv. et Val.* Hist. des Poiss. tom. iii. p. 357.

FORM.—Very much resembling that of the *U. trifasciatus,* but with the following differences. The eyes rather larger, distant from the end of the snout rather more than two diameters and a half. Suborbitals traversed towards their lower margins by a number of lines, each terminating in a pore, and with their whole disks studded besides with pores without lines: the lower margin itself presents four distinct deeply-cut notches, the first of which receives the end of the maxillary when the mouth is closed. A single row of small conical teeth in each jaw ; in addition to which, in the upper, there are some stronger ones in front, exterior to the others, amounting to eight in number, the central pair of which bends inwards or towards each other, and the three on each side, which are the strongest of all, backwards and outwards. No teeth on the vomer or palatines. The posterior extremity of the maxillary is much narrower than in the last species. Spine of the opercle sharp and well developed, about two lines and a quarter in length. Barbules reaching very nearly to the posterior margin of the opercle. Ramifications of the mucous tubes on the lateral line very numerous. Height of the first dorsal equalling rather more than half the depth. Space between the two dorsals equalling half the length of the second dorsal. This last fin pointed behind, as well as the anal, but not so much so as in the *U. trifasciatus.* Pectorals when laid back reaching to a vertical line from the extremity of the first dorsal. Ventrals reaching a little beyond the pectorals, but falling short of the anal by a space equalling half their own length.

D. 8—9; A. 7; C. 15, &c.; P. 16; V. 1/5.

Length 8 inches.

COLOUR.—"Vermilion, with streaks of iridescent blue."—D. In spirits, the colour appears of a uniform dull reddish yellow, without any indication of spots or other markings on the fins or body.

Habitat, Porto Praya, Cape Verde Islands.

I suppose this to be the *U. Prayensis* of Cuvier and Valenciennes, the description of which, so far as given in the "Histoire des Poissons," is tolerably applicable. Those authors, however, mention a spot in the middle of each scale of a deeper red than the ground colour, which is not alluded to by Mr. Darwin in his notes, and of which I see no trace on the fish in its present state. On the other hand they are silent with regard to the blue streaks. In some of its characters, especially as regards the teeth, this species seems to approach the *U. maculatus*; but the colours are different in this last also, which is moreover found on the opposite side of the Atlantic.

FAMILY.—TRIGLIDÆ.

TRIGLA KUMU. *Less. et Garn.*

Trigla kumu, *Less. et Garn.* Zoologie de la Coquille, (Poissons) Pl. 19.
——————— *Cuv. et Val.* Hist. des Poiss. tom. iv. p. 36.

FORM.—In general appearance very much resembling the *T. Hirundo*, but more elongated. Depth contained about five times and a half in the entire length. Head rather more than four times and a quarter in the same. The obliquity of the profile about the same as that of the *T. Hirundo*, but the concavity of the interocular space less. The granulations on the head not so coarse, or so strongly marked, the lines in which they are arranged being closer and more numerous: those on the suborbitals radiate from a point nearer the extremity of the snout: no crest or ridge at the bottom of the suborbital, and only a very indistinct one at the bottom of the preopercle: as Cuvier has well noted, the grains on the border of the preopercle are divided into little isles, or collected in clusters, by irregular lines which undulate amongst them; and in this specimen, the same character presents itself on the posterior and upper portion of the suborbital: some of the first lines on the opercle are plain, or without granulations. Snout emarginated, with three or four denticulations on each side rather sharper and more developed than in the *T. Hirundo*. Two spines at the anterior angle of the eye; but none at the posterior angle, or on the temples. Suprascapular, opercular, and clavicular spines much as in the *T. Hirundo*. Lateral line and whole body smooth, excepting the dorsal ridges, which are strongly serrated. Dorsal spines as in the *T. Hirundo;* second longest; the first with a series of obsolete granulations on its anterior edge. Pectorals not quite one-third of the entire length: free rays incrassated in the middle, tapering towards the ends, but with the extreme tips slightly dilated.

D. 10—16; A. 16; C. 11, &c.; P. 11, and 3 free; V. 1/5.

Length 16 inches 6 lines.

Colour.—" Whole body bright red."—D. The pectorals, of which no note was taken in the recent state, appear, in the dried skin, externally, of a dusky colour, approaching to black, with white rays; the lower margin, however, is paler, and was probably originally like the body: inside, the colour is much the same, but variegated with a few white spots; there are also portions of a paler tint, probably the remains of a fine blue. I see no distinct trace of the large deep black spot, said by Cuvier to occupy the seventh and eighth rays on the posterior face of the fin.

Taken in the Bay of Islands, New Zealand. The only respect in which it differs from the description of the *T. kumu* by Cuvier and Valenciennes, is its having one more spine in the first dorsal.

1. PRIONOTUS PUNCTATUS. *Cuv. et Val.*

Prionotus punctatus, *Cuv. et Val.* Hist. des Poiss. tom. iv. p. 68.

Form.—Well characterized by the form of the snout, which is very obtuse, and as it were truncated, with scarcely any notch in the middle; the margins of the lobes are crenated with minute denticles, immediately beyond which is a small sharp spine directed backwards; further on, almost immediately above the corners of the mouth, is a second similar, but somewhat larger spine. There are also some minute spines on the temples, as well as on the ridge of the preopercle, besides the ordinary spines, common to other species, which in this are all well developed and very sharp. Dorsal spines smooth, or only the first with a faintly marked line of granulations; third longest. Pectorals long, contained exactly two and a half times in the entire length; when laid back, they reach to within two rays of the extremity of the second dorsal. Free rays rather slender and tapering, with the extreme tips pointed; not above half the length of the pectorals. Ventrals rather longer than the free rays.

D. 10—12; A. 11; C. 11, &c.; P. 13 and 3; V. 1 5.

Length 8 inches 9 lines.

Colour.—" Above and sides olive brown, with red spots and marks; beneath silvery white; edges of the pectoral fins Prussian blue."—D.

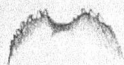

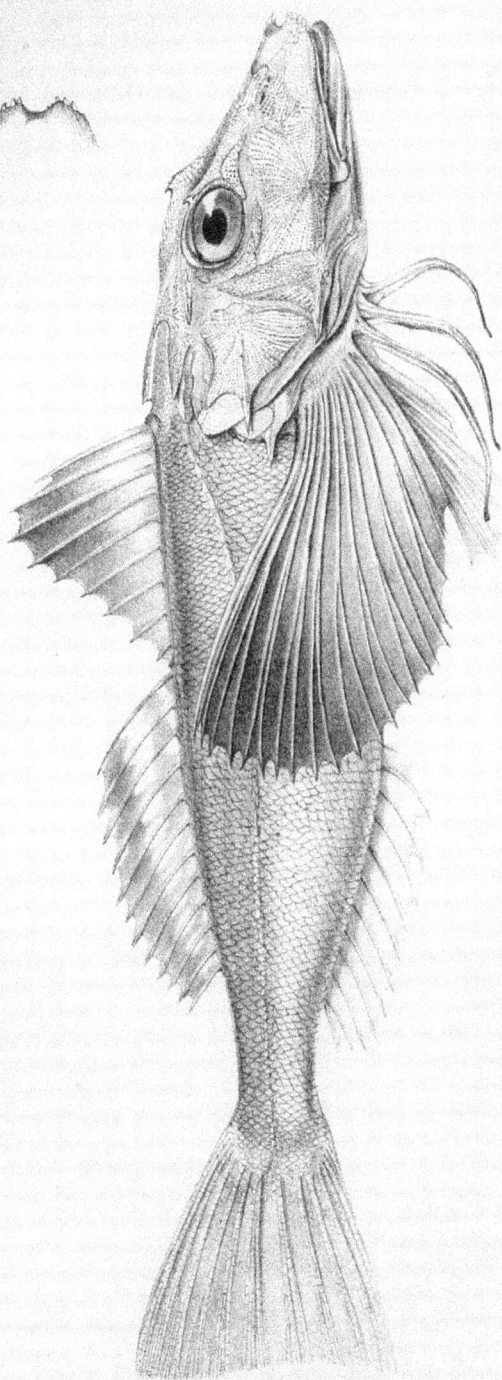

Prionotus Miles. *Vaillant.*

This species is said by Cuvier and Valenciennes to be common all along the Brazilian coast as far as the mouth of the Plata. Mr. Darwin's specimen was taken swimming on the surface in the Bay of Rio de Janeiro, and agrees well with the description by those authors. " When first taken made a croaking noise."—D.

2. PRIONOTUS MILES. *Jen.*

PLATE VI.

P. splendidè rubro variatus; rostro emarginato, utrinque distinctè denticulato ; buccis levissimè granulosis; fossulá dorsali lateribus inermibus ; squamis corporis parvis, ubique ciliatis; pinnis pectoralibus modicis, corpore certè triplò brevioribus ; radiis liberis subincrassatis, apicibus dilatatis.

B. 7 ; D. 10—12; A. 11 ; C. 12, &c.; P. 13 et 3; V. 1/5.

Long. unc. 10. lin. 3.

FORM.—In general form, that of the head especially, very similar to the *Trigla Hirundo* of the British seas. Compared with the *P. punctatus* last described, it is rather more elongated, the depth and thickness being less. Profile falling less obliquely. Space between the eyes broader, but equally concave. Snout not so obtuse, and more deeply notched; with six short but well developed teeth on each side, followed by some minuter denticles. The lines of granulations on the snout and cheeks are very fine, and not so strongly marked, or spread over so large a portion of the face. One principal spine, preceded by two or three small denticles, at the anterior angle of each orbit ; at the posterior angle, a well marked notch with a small denticle, (in this specimen the denticle on the left side of the head only,) but no regular spine : these notches are connected by a transverse line on the cranium, but not by a groove (as in *P. Carolinus*, Cuv. et Val.). No spines on the temples, or on the crest at the bottom of the preopercle ; but the ordinary spine of the preopercle, as well as the opercular, suprascapular, and clavicular spines, appear as usual, though not quite so long as in the *P. punctatus;* the clavicular spine has one line of points along its ridge, but the granulations are not very obvious. Band of palatine teeth much as in *P. punctatus.* First dorsal spine with a row of granulated points in front; the second spine with a row on the left side of the fin ; the third spine with a very rudimentary row on the right side; but none of these granulations very obvious: third spine longest, equalling about three-fourths of the depth of the body; the fourth and succeeding spines gradually decreasing to the tenth, which is barely visible, and so reclined as to be easily overlooked. Dorsal groove shallow, with the sides unarmed. Scales on the body small, broader than long ; their free edges finely ciliated, communicating a decided roughness to the touch ; their concealed portions crenated at the hinder margin, and marked with a fan of five or six striæ. Lateral line not distinguished by any particular scales, but forming a whitish streak from the upper part of the gill-opening to the caudal. Pectorals relatively shorter than in *P. punctatus,* contained a little more than three times in the entire length; when laid back they reach to a vertical line from the fourth

ray of the second dorsal. Free rays rather stout, with their tips somewhat dilated and approaching to spatuliform ; in length about two-thirds that of the pectorals.   Ventrals a trifle
longer than the first or longest of the free rays.

COLOUR.—" Above mottled brilliant tile red ; beneath silvery white."—D.   Mr. Darwin is rather
doubtful whether by the above description, he meant that the entire fish was brilliant red, or
only mottled with red upon some obscure ground.

Habitat, Galapagos Archipelago.

Taken at Chatham Island, in the Galapagos Archipelago, and decidedly distinct from all the species described by Cuvier and Valenciennes.   From *P.
strigatus* it differs in the finer granulations of the cheeks, less obtuse and more
deeply notched snout, smooth scales, and absence of a second lateral line ; from
P. Carolinus in the want of a transverse groove on the cranium, and in the fin-ray
formula, but it resembles this species in the dilated tips of the free rays ; from
P. punctatus as pointed out in the description ; from *P. tribulus* in the want of
the spine on the suborbital, and in its much shorter pectorals.   These fins indeed
are shorter than in any of the above-mentioned.

As all the species described in the " Histoire des Poissons," are found on
the Atlantic side of America, the geographical range of this genus is extended
to the Pacific by the discovery of the present one.

FAMILY—COTTIDÆ.

ASPIDOPHORUS CHILOENSIS.   *Jen.*

PLATE VII. FIG. 1.   Lateral view twice nat. size.
Fig. 1*a*. Dorsal view nat. size.
Fig. 1*b*. Lateral view nat. size.

*A. corpore elongato, antìcè octagono, posticè hexagono; vomere et ossibus palatinis
dentibus distinctis instructis; maxillis subæqualibus; rostro ultrà fauces haud producto; mento et membranâ branchiali cirratis: pinnis dorsalibus discretis; primâ
radiis gracilibus.*

B. 6; D. 8—7; A. 8; C. 11⅔; P. 14; V. 1/2.

Long. unc. 2. lin. 7.

FORM.—More elongated than the *A. cataphractus*, which it somewhat resembles in general appearance.   Anterior portion of the body octagonal, and the posterior, or all beyond the second
dorsal and anal, hexagonal.   Head equally depressed as in that species; but its breadth less,
being only one-fifth of the entire length, caudal excluded.   Length of the head rather less than

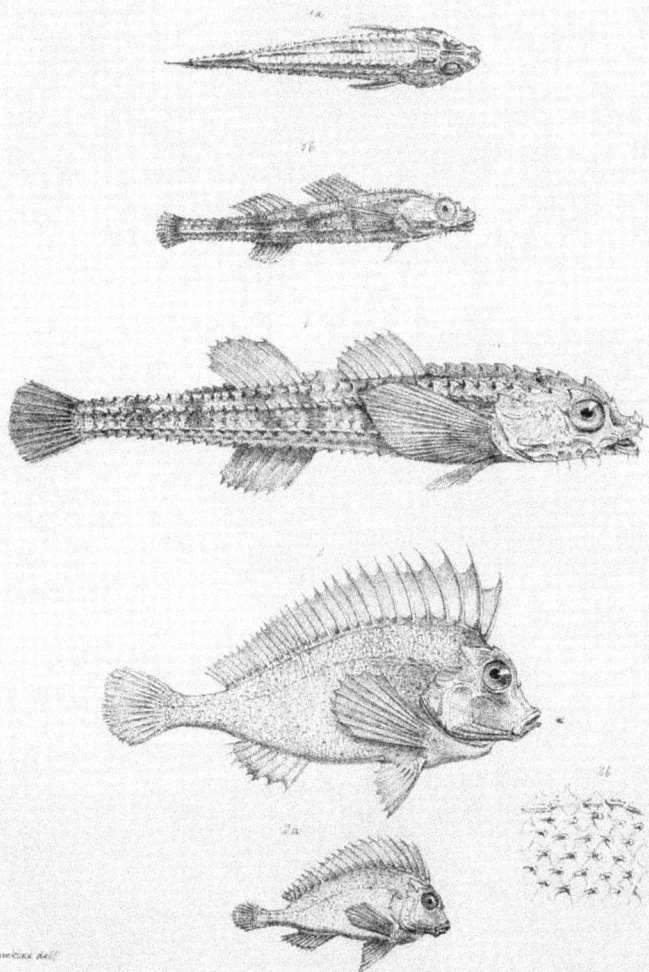

N. Hawkins delt.

1. *Aspidophorus Chiloensis* Twice Nat. Size.
1a. 1b. Nat. Size.
2. *Agriopus hispidus.* Twice Nat. Size.
2a. Nat. Size.
2b. Magnified Scales.

one-fourth of the entire length. Depth at the nape rather less than one-seventh of the same. Eyes relatively a little larger than in *A. cataphractus ;* their diameter one-fourth the length of the head ; placed high in the cheeks, and distant one diameter from the end of the snout. Upper part of the orbit elevated into an osseous ridge on each side of the crown of the head, with a spine at its anterior angle, and the ridge itself terminating in a sharp, rather stronger, spine at the posterior angle ; both spines directed backwards. Space between the eyes concave, equalling in breadth not quite one diameter of the eye, with two longitudinal sharp ridges running respectively parallel to the ridges of the orbits, but not nearly so much elevated as these last ; these ridges terminate posteriorly at a groove, which runs transversely behind the eyes, separating the vertex from the occiput. The snout presents the same four spines, which are to be seen in the *A. cataphractus,* but it does not project beyond the mouth. The lower margin of the suborbital presents a somewhat irregular ridge formed by a series of bluntish tubercles, the last of which terminates in a very minute spine directed backwards. Limb of the preopercle with three diverging smooth ridges, dilating at their extremities into three flattened blunt points, which project a little beyond the membrane, but can scarcely be called spines. Opercle with one ridge not so strongly marked as those of the preopercle, and not terminating in any distinct point, nor even reaching quite to the edge of the membrane. Jaws nearly equal; but the upper one a very little the longest ; each with a narrow band of minute velutine teeth : a distinct chevron of similar teeth on the front of the vomer, and a short imperfect row on each palatine. Tongue smooth. Gill opening large : the branchial membrane not notched, but passing transversely over the isthmus, to the edge of which it is nevertheless attached on each side. Chin clothed with short fleshy cirri ; also a few on the lower jaw and branchial membrane ; but they are much shorter, and less conspicuous than in the *A. cataphractus,* especially on the branchial membrane, where they are very sparingly scattered. The occiput presents the four usual ridges formed of granulated tubercles ; and between the innermost pair there is also a much less conspicuous, but slightly raised line running longitudinally down the middle : the two innermost of the above ridges are nearly in a line respectively with the two ridges of the orbit, behind which they commence, and they would pass on to unite with the two dorsal carinæ were they not separated from the latter by a deep transverse depression at the nape : the two outermost of the occipital ridges commence behind the eyes themselves, and terminate at the suprascapulars, each in a sharp point directed backwards, but not prolonged into a spine. The carinated scales which arm the body of this species, are more sharply serrated than those of the *A. cataphractus,* the keels terminating behind in hooked points; and the elevated lines which form the striæ on each side of the keel are fewer in number and more raised. The ridges which they form are also more marked, and the second ridge on each side commences immediately behind the angle of the opercle, instead of opposite the vent as in that species ; so that the whole body is perfectly octagonal from the gills to the termination of the dorsal and anal fins :* at that point, the two dorsal ridges and the two ventral unite respectively to form one, or rather approximate so closely as to form but one in appearance ; for, if closely examined, there will still be found two parallel rows of serratures. In each of the two uppermost or dorsal ridges, there are twenty-seven scales, reckoning from the hollow at the nape to the point where the ridges unite. In the second ridge (which extends, as before observed, from the gills to the caudal) there are thirty-

---

* In the *A. cataphractus,* the body is hexagonal from the gills to a little beyond the vent; octagonal from this last point to the termination of the dorsal and anal fins ; then hexagonal again to the end of the tail.

eight. In the third, which commences behind the pectoral, and extends in like manner to the
caudal, there are thirty-five. In the fourth, which commences on the breast, immediately behind
the point of attachment of the branchial membrane to the isthmus, there are thirty, reckoning
to behind the anal, where it unites with its fellow to form one; between this point and the
caudal there are ten, the serratures of which are rather obsolete. The fourth pair of ridges
are throughout their course less sharply serrated than the second and third pairs, and these last
again rather less so than the first or dorsal pair. Between the two ventral ridges, near their
commencement in front of the ventral fins, are six slightly serrated scales (similar to those in
the ridges) forming on the breast a somewhat triangular patch, two single ones standing first,
then four others in pairs. The lateral line, which is catenulated as in *A. cataphractus*, commences at the upper angle of the opercle, then bends downwards to take a middle course between
the second and third ridges, which it preserves to the caudal. The first dorsal commences
behind the seventh scale in the dorsal ridges, or at about one-third of the entire length; it is of
the same form as in the *A. cataphractus*, but contains more rays; its membrane terminates at
the fifteenth scale, and there are rather more than two scales between it and the second dorsal,
which last is rather shorter and higher than the first. The rays of the first dorsal are not stouter
than those of the second, nor relatively stouter than those of the *A. cataphractus*. The rays
of the second dorsal are simple, with the second and third rather longer than the first. The
anal answers to the second dorsal. The pectorals are rounded, and one-fifth of the entire
length. Ventrals very narrow, and scarcely more than half the length of the pectorals.
Position of the vent a little anterior to a line connecting the extremities of the ventrals.

Colour.—(*In spirits.*) Dusky grey above and on the sides, paler beneath; with four broad transverse blackish fasciæ passing across the back and down the sides as far as the third longitudinal
ridge of scales. The first fascia is in the region of the first half of the first dorsal; the second
at the commencement of the second dorsal; the third near the end of the second dorsal; the
fourth half way between the end of the second dorsal and the caudal; and a little beyond this
there is a faint trace of a fifth fascia. The body is a little mottled in places with spots of the
same dark colour as the fasciæ, and the fins, with the exception of the ventrals, are of the same
hue.

Habitat, Chiloe, (West coast of S. America).

The absence of vomerine teeth has been considered by Cuvier as one of the
characters serving to distinguish *Aspidophorus* from *Cottus;* but as these teeth
are very distinctly developed in the present species, we must rather dwell upon
the large keeled sharp-pointed scales, which envelope the body in a kind of mail,
and, as Dr. Richardson observes,* " give the *Aspidophori* a totally different aspect
from the *Cotti*." Indeed on equally strong grounds as those on which Cuvier has
separated *Pinguipes* from *Percis* and *Priouotus* from *Trigla*, the present species,
which possesses both vomerine and palatine teeth,† might be made a distinct

---

* Faun. Bor. Amer. Part Third, p. 49.

† Is it not possible that this may be found to be also the case with several of the foreign species described by
Cuvier, in which the absence of these teeth has been rather presumed than ascertained from actual examination?

genus from *Aspidophorus*, or at least considered as one of its subgenera. But in the present uncertain state of our knowledge with respect to the exact value of this character,* and from the general resemblance of the *A. Chiloensis* in all its principal characters to the other species of this genus,† I have not thought this step necessary.

This species was taken by Mr. Darwin at Chiloe. There are two specimens in the collection. The second differs from the one above described, only in having one ray less in the first dorsal, and two more carinated scales in each of the dorsal ridges. Independently of its having vomerine and palatine teeth as above noticed, this species will not enter into any of Cuvier's sections of the genus *Aspidophorus*, but combines in itself the characters of his first and third ; the dorsals being separated by nearly three scales, the jaws being very nearly equal, the rays of the first dorsal not stouter than those of the second, and the throat being bearded.

PLATYCEPHALUS INOPS. *Jen.*

P. capite longo, lævi, ubique inermi, spinis duábus ad angulum preoperculi brevissimis æqualibus exceptis; oculis magnis, arctè propinquantibus: dorso et lateribus fuscis; abdomine albido ; pinná dorsali primá liturá magná irregulari nigro-fuscá posticè maculatá ; dorsali secundá, cauduli, et pectoralibus, maculis fuscis parvis; anali et ventralibus ferè omnino nigricantibus.

B. 7; D. 8—12 ; A. 12 ; C. 13, &c. ; P. 19; V. 1/5.

LONG. unc. 16.

FORM.—Head very much depressed, and rather longer than in most of the species of this genus ; its length being nearly twice its own breadth, and nearly one-third of the entire length. Breadth of the body at the pectorals one-seventh of the entire length : depth at that point half the breadth. Snout rounded horizontally. Lower jaw longest. Gape reaching to beneath the

---

* Cuvier seems to have attached much value to the character of teeth on the palate; but I agree with Dr. Richardson, (*Faun. Bor. Am.* Part iii. p. 19.) in considering it " of little importance as a generic character in some families of fish." And the author last mentioned notices an instance (exactly analogous to that of the *Aspidophorus Chiloensis*) in the *Thymallus signifer*, which, he says, " resembles the common grayling very closely in its general form, but differs from it in having palatine teeth."

† In its general characters it does not depart from the *A. cataphractus* of the British seas, anything like so much as the *A. quadricornis*, and *A. monopterygius* do.

F

anterior margin of the orbit. A band of sharp velutine teeth in each jaw broadest above; a double semicircular patch of similar teeth on the front of the vomer, and a band all along each palatine as broad as that in the lower jaw. Branchial arches and pharyngeans rough with similar teeth. Tongue free, thin, flat, truncated at the apex with a double emargination in the middle, of equal breadth throughout, without teeth, the central portion cartilaginous with a broad membranous border all round. Eyes large, their diameter one-sixth the length of the head, approximating, with not half a diameter between, distant two diameters from the extremity of the lower jaw. The spines of the preopercle (which in some species are long and very unequal) very short and inconspicuous, of equal length, the lower one rounded off almost to nothing. Head smooth all over; presenting the usual ridges, which however are not very salient, but with hardly anything deserving the name of spines, excepting only a small flat spine terminating the opercle, and a minute but sharp one on the upper ridge of the scapula: none at the anterior angle of the first suborbital, or on the ridge of the orbit. Gill opening very large; the branchial membrane notched underneath for its whole length.

Pectorals broad and oval but short, contained nearly eight times in the entire length; the first two rays simple, the next ten branched, the last seven, which are rather stout, again simple. Ventrals separate by nearly the whole breadth of the body, attached beneath the middle of the pectorals, longer than these last fins by nearly one-third, and reaching very nearly but not quite to the vent, which is a little posterior to the middle of the entire length: the spine of the ventrals is one-third of the longest of the articulated rays which are the last or innermost. The first dorsal commences above the middle of the pectorals, and occupies between one-sixth and one-seventh of the entire length; its greatest height is about two-thirds of its own length; the first spine is very short, and detached, as in the other species; the second a little shorter than the third which is longest; the rest gradually decrease to the last, which is one-third the length of the second; this fin therefore is not so triangular as in many of this genus. A small space between the first and second dorsals. This last longer and rather lower than the former, contained four and a half times in the entire length; all the rays nearly even, with the exception of the first only, which is a little shorter than the second. Caudal square. The anal answers to the second dorsal, but begins, as well as terminates, a little backwarder.

The lateral line commences at the suprascapular, and gradually bends down till it reaches the middle of the depth which it keeps for the remainder of its course; it is perfectly smooth throughout. The scales cover all the body and a part of the head, but are not present between the eyes, or on the front of the snout, or on the jaws. They are small, oblong-oval, finely striated, with a fan of eleven or twelve deeper striæ posteriorly, their free edges cut square, not ciliated.

COLOUR.—(*In spirits.*)—Back and sides nearly uniform deep brown; beneath white; the two colours separated by a well-defined line. First dorsal transparent, with a deep brown stain or blotch on the membrane, of an irregular form, and occupying more than the posterior half of the fin. Second dorsal uniformly, but rather obscurely, spotted throughout. Caudal with transverse rows of similar spots. Anal nearly uniform pale dusky, the spots hardly distinguishable from the ground. Ventrals the same. Pectorals with spots on the rays, but with the intervening membrane nearly transparent.

Habitat, King George's Sound, New Holland.

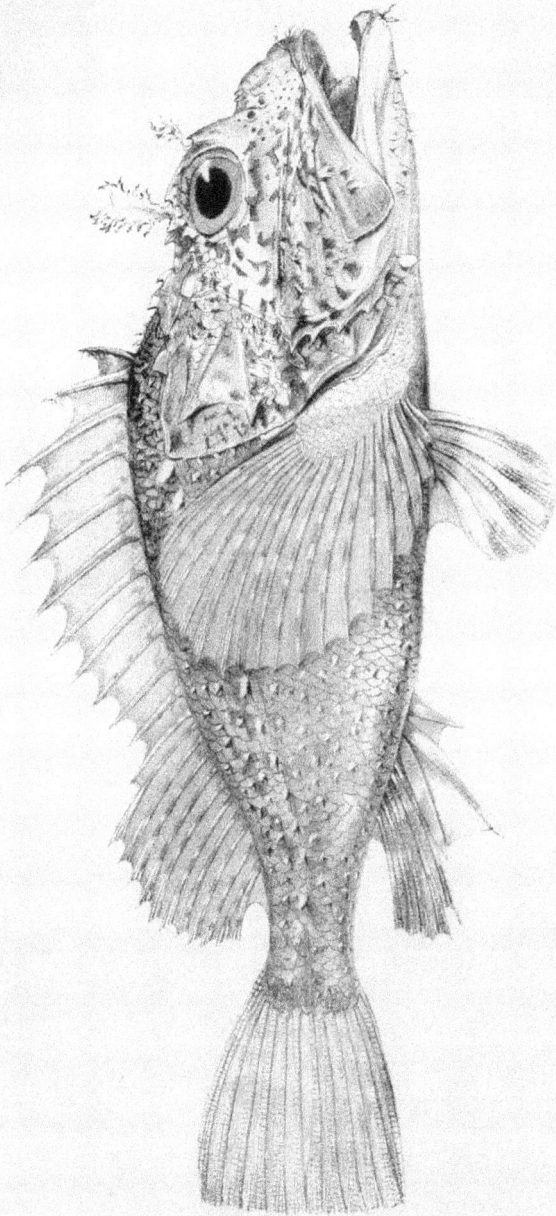

Scorpaena Histrio *Nat. Size*

This species very closely approaches the *P. lævigatus* of Cuvier and Valenciennes, with which it particularly agrees in the smoothness of its head, and large approximating eyes. The two spines, however, at the angle of the preopercle appear to be still smaller than in that species;* the fin-ray formula is a little different; and so also are the colours; the first dorsal being particularly characterized by a large irregular dark-coloured stain on its posterior portion, and the anal and ventrals being almost wholly dusky, instead of pale with spots on the rays only, as in the *P. lævigatus.* Possibly it may be a mere variety. Mr. Darwin's specimen was obtained at King George's Sound.

Family—SCORPÆNIDÆ.

Scorpæna histrio. *Jen.*

Plate VIII.

S. toto corpore coccineo, pinnis pallidioribus, maculis parvis irregularibus nigricantibus: capite magná ex parte alepidoto, lineis spinosis solitis armato: pinná dorsali spinis inæqualibus, tertiá paulo longissimá: capite et lateribus cirris cutaneis parvis ubique adornatis; quatuor palpebralibus, præsertim duobus posterioribus, majoribus, palmatis.

B. 7; D. 12/9; A. 3/5; C. 13, &c.; P. 20; V. 1/5.

Long. unc. 9.

Form.—General form resembling that of the *S. Scrofa.* Depth at the pectorals just one-fourth of the entire length. Thickness a trifle more than two-thirds of the depth. Head more than one-third of the entire length. Eyes large and elevated, distant from the end of the snout rather more than one diameter; the space between very concave, twice as long as broad, with two whitish lines in the central furrow, diverging as they recede backwards towards the nape, but scarcely elevated into salient ridges. Mouth oblique, with the gape large and the lower jaw a little the longest; when closed, the end of the maxillary, which is broad and much dilated, reaches to a vertical line from the posterior part of the orbit. A broadish band of velutine teeth in each jaw as well as on the vomer and palatines. Tongue smooth. A small sharp triangular spine on each of the nasal bones, (in this specimen that on the left side is double or forked): upper margin of the orbit, which is much elevated, with three spines, one strong one at the anterior angle, and two, nearly as large, further back; beyond which, on the left orbit only, is a fourth smaller one. Space between the eyes bounded posteriorly by a raised arc

---

* Judging from the figure in the "Voyage de l'Astrolobe (Zoologie)," pl. 10. f. 4.

having the curvature inwards, with a spine on each side ; this is followed by the depressed occiput, which forms a hollow ; and on each side of this, at its posterior margin, or at the commencement of the nape, are two other strong spines : there are likewise two spines at the suprascapulars, and between these and the posterior margin of the orbit of the eye, on what may be called the temples, are two more ; of these last, the first, which is small and close to the orbit, is double ; the second, which is larger and situate a little above the upper angle of the preopercle, is, in this specimen, double on the right side and single on the left. The first suborbital has two spines on its anterior margin, the first directed forwards, and the second downwards ; on its disk are two salient ridges, which are unarmed, and not very conspicuous. The second suborbital is entirely without spines, but elevated in the middle into a double smooth ridge or crest. Margin of the preopercle with six spines ; the second longest ; the first, as well as the two lowermost, small and inconspicuous. Opercle with two osseous diverging ridges terminating in spines : the scapular and clavicular bones likewise terminate each in a flattened spine. Lateral line and scales much as in *S. Scrofa* ; the latter with their free edges perfectly smooth.

The cutaneous filaments and appendages on this species are as follows : three small ones at the extremity of the snout ; one small but broad one at the upper margin of the anterior orifice of the nostril ; two very conspicuous palmated ones on each orbit, especially the posterior one, which is largest, and very broad ; two on the margin of the first suborbital ; some small ones on the cheeks and maxillaries ; six beneath the lower jaw, two being near the symphysis, and two on each ramus ; a row on the margin of the preopercle, and very numerous small ones scattered about the nape and sides of the body, of which a row along the lateral line are rather more conspicuous than the others.

The spines of the dorsal fin are moderately strong, and unequal ; the first is rather more than half the length of the second, which itself is two-thirds of the third ; this last is less than half, but more than two-fifths, of the depth of the body ; from the third, the spines decrease very gradually to the eleventh, which is a little longer than the first ; the twelfth is higher than the eleventh by one-third : soft portion of the fin rounded, and where most elevated just equalling in height the third or longest spine. Anal spines very unequal ; the first not very stout, and less than half the length of the second, which is very strong indeed, as well as the longest of the three ; the third is stouter than the first, but not nearly so stout as the second, though nearly equalling that spine in length : soft portion of this fin with its greatest elevation rather exceeding the second spine. Caudal slightly rounded. Pectorals rather more than one-fifth of the entire length ; the ten lowermost rays simple ; the nine immediately above these branched ; the uppermost of all simple like the bottom ones, but slenderer as well as shorter than the others. Ventrals not above two-thirds the length of the pectorals ; in other respects as in *S. Scrofa*.

A second specimen.—Smaller than the one above described, measuring seven inches and a half in length. The two diverging lines on the cranium between the eyes are rather more salient, and the left orbit is without the fourth spine ; but in all other respects, including the fin-ray formula, the two specimens are exactly similar.

COLOUR.—" Whole body scarlet red, fins rather paler ; with small irregularly-shaped light black spots."—D.

Habitat, Chatham Island, Galapagos Archipelago.

FISH.   37

This species differs more or less in the details of form, as well as colours, from all those which I can find described by authors. Most of the foreign species of this genus noticed by Cuvier and Valenciennes, come either from the eastern coast of America or the East Indies; and they do not appear to have received any from that part of the Pacific, whence the present one was obtained.

SEBASTES OCULATA. Val.?

Sebastes oculata, Cuv. et Val. Hist. des Poiss. tom. ix. p. 344.

FORM.—Greatest depth contained about three times and three quarters in the entire length. Head about one-third of the same. Eyes large; the interocular space, equalling rather more than half their diameter, concave, with two longitudinal ridges running respectively parallel to the two superciliary ridges. Two spines on the upper part of the snout, in a line with the nostrils; one at the anterior part of the orbit; three at the posterior, passing off in a line towards the occiput, where there are two other moderately strong ones terminating the lateral occipital ridges; five very strong spines or teeth edging the rounded angle of the preopercle; two sharp ones at the posterior angle of the opercle, the upper one most developed; one at the scapula, and two at the suprascapular. There are either three orifices to each nostril, or else, adjoining the two usual openings, a large pore so manifest (at least in this specimen in its dried state) as easily to be mistaken for a third: this additional one is close to the nasal spine. Dorsal spines of only moderate strength: anal stronger, especially the second, which is very stout, as well as the longest of the three; the third, however, is more than half the length of the soft rays. Pectorals broad and rounded; their length contained about four times and a half in the entire length; first ray simple, the next eight branched, the nine lowermost simple again, and rather stouter than the others. The caudal appears to have been square. Scales small and ciliated, covering nearly the entire head, as well as body, but very thinly scattered on the extremity of the snout in advance of the nostrils.

D. 13/14; A. 3/6; C. 14, and 3 shorter ones; P. 18; V. 1/5.

Length 10 inches.

COLOUR.—" Under surface, sides, branchial covering, and part of the fins, ' tile and carmine red;' dorsal scales pale yellowish dirty brown."—D. In its present dried state, the colour is of a uniform brown.

Habitat, Valparaiso.

This species is probably the <i>S. oculata</i> of Valenciennes; but the depth rather exceeds, and in its recent state it must have still more exceeded, one-fourth of the entire length, the proportion given in the " Histoire des Poissons." The spines on the opercle and suprascapular also can hardly be called " smaller," as there stated, than those on the orbit and occiput, at least the upper one on the opercle. There are also two soft rays less in the anal. It may be added further, that Mr. Darwin's notes make no mention of the four brilliant rose-coloured spots along the base of the dorsal fin, spoken of by Valenciennes.

The *S. oculata* was discovered by M. Gay at Valparaiso, where Mr. Darwin's specimen also was obtained. It is the only species of this genus on record brought hitherto from South America. It may be stated, however, that Mr. Darwin has a drawing of another species, made by Mr. P. King, found also at Valparaiso, differing from the above in having the spines on the head less developed, and apparently, in some of its characters, approaching the *S. variabilis*. This last is a species inhabiting the sea which separates N. America from Kamtschatka.

AGRIOPUS HISPIDUS. *Jen.*

PLATE VII. FIG. 2.   Twice nat. size.

Fig. 2 *a.* Nat. size.

Fig. 2 *b.* Portion of the hispid cuticle magnified.*

A. pallidè rubro-aurantius, dorso nigricante, pinnis nigro-maculatis: corpore hispido, altitudine tertiam partem longitudinis æquante'; spinis nasalibus duábus parvis recurvis ; vomere dentibus velutinis minutissimis instructo : pinnâ dorsali inæquali, anticè allevatâ, spinis quartâ et quintâ paulo longissimis, succedentibus gradatim diminutis, ultimâ radiis articulatis multò breviori.

D. 17 13; A. 1 8; C. 13, &c.; P. 9; V. 1/5.

LONG. unc. 1. lin. 9.

FORM.—General form resembling that of the *A. torvus,* but the depth much greater, equalling one-third of the entire length, or very nearly. Length of the head somewhat less than the depth of the body. The line of greatest depth passes through the insertion of the pectorals. The profile viewed apart from the superciliary ridges, which are sharp and prominent, falls in a straight but very oblique line from the commencement of the dorsal to the mouth. On each side of the median line of the snout, in advance of the eyes, is a small but sharp spine, directed upwards and backwards. There are also two minute spines on the first suborbital immediately above and behind the end of the maxillary ; these are placed one over the other, the uppermost, which is the sharpest and most conspicuous, taking an upward direction like the

* Called by mistake in the plate " magnified scales."

nasal spines, the lowermost, which is blunt and not so obvious, a downward one. Mouth small, without any teeth that can be discerned even with a lens; but a decided roughness can be felt on the vomer, seeming to indicate the presence of minute teeth on that part. The superciliary ridges, already alluded to, are slightly granulated, and terminate behind in two sharp triangular points. The occipital ridges, a ridge on the posterior suborbital immediately beneath the eye, and an interrupted ridge on the temples and suprascapulars, are in like manner granulated, or rather obscurely crenated. The opercle and preopercle are marked with a few striæ, but show neither granulations nor spines. Gill-opening very small. No scales on any part of the head and body; but the whole surface of the latter is hispid with minute bristly appendages to the cuticle, each springing from a minute papilla. There are also a number of fine lines traversing the cuticle in two directions, and forming a kind of net-work. The lateral line commences at the suprascapular, and terminates a little beyond the end of the dorsal, not reaching quite to the caudal; its course is nearly, but not exactly, parallel to the dorsal line, the distance between them being at first one-third, but towards the caudal between one-third and one-fourth of the depth.

Dorsal very much elevated anteriorly, but its height by no means uniform throughout; the first spine one-fourth shorter than the second; this again a little shorter than the third; and this last a very little shorter than the fourth and fifth, which are longest, and which equal three-fourths of the depth of the body; sixth and succeeding ones gradually decreasing, the ninth being about equal to the first, the twelfth about one-third shorter; the next four are scarcely shorter than the twelfth, and the seventeenth or last is a little higher than the sixteenth; then follows the soft portion of the fin, which is here again elevated, the soft rays being nearly double the length of the last spinous.* The anal answers in position to the first two-thirds of the soft dorsal, terminating before that fin, as in *A. torvus :* the fourth, fifth, and sixth soft rays are longest, and much longer than the soft rays of the dorsal; the spine is short and slender, and not much more than half the length of the first soft ray. The last ray of both dorsal and anal is divided quite to the root so as to appear as two. The caudal appears to have been rounded, but the ends of the rays are worn and broken. Pectorals long, equalling one-third of the entire length : they consist of nine rays, the three middle ones of which are longest; the three upper and the three lower ones are respectively equal; all the rays simple. Ventrals much shorter than the pectorals, and, though attached rather more behind, not reaching so far; their spine is rather stout, much more so than that of the anal, and about three-fourths the length of the first two soft rays, which are the longest in the fin.

COLOUR.—" Pale reddish orange, with black spots on the fins, and a dusky shade on the back."—D.

A second specimen only differs from the above in having the teeth in the jaws more sensible to the touch, though still scarcely to be seen; and in the superciliary and occipital ridges being less granulated or crenated at the edges. The colours also are a little darker. The fin-ray formula is exactly the same in both specimens.

Habitat, Peninsula of Tres Montes, Archipelago of Chiloe.

This species approaches most nearly the *A. Peruvianus* of Cuvier and Valenciennes, with which it agrees in the great depth of the body, and in the

---

* This portion of the fin is not quite correctly represented in the plate, being made too low, in consequence of the rays having been broken at their extremities in the specimen figured.

presence of two nasal spines; but it differs in the roughness of the skin (that species being described as smooth), and in the greater inequality of the dorsal fin. Perhaps it may be the same as the species brought from the coast of Chili by Mr. Cuming, and briefly noticed by Mr. Bennett in the "Proceedings of the Zoological Society" (1832, p. 5.), but which this last gentleman did not venture to describe as new, from the circumstance of its general agreement with the *A. Peruvianus*. The principal deviation in Mr. Cuming's fish from the species just mentioned is stated to have occurred in the number of the fin-rays; those of the spinous portion of the dorsal fin being seventeen (one less than in the *A. Peruvianus*), while of the soft rays of the anal there were ten (three more than in the species referred to). Mr. Darwin's fish agrees with Mr. Cuming's in the number of the dorsal spines, but not in that of the soft rays of the anal, which is eight, being one more than in the *A. Peruvianus* and two less than in Mr. Cuming's; and it is observable that both the specimens obtained by Mr. Darwin agree in this particular. Mr. Bennett has not noticed any of the other characters of Mr. Cuming's fish.*

One of the most distinguishing peculiarities in the species here described is the existence of vomerine teeth, which though extremely minute are quite sensible to the touch. As these teeth are denied by Cuvier to the whole genus, we have here another instance, similar to that of the *Aspidophorus Chiloensis* already mentioned, of the slight value of the character which their presence or absence affords. Possibly, however, they may disappear in the adult state. Both Mr. Darwin's specimens are small, neither equalling two inches; and if they are immature, which is probably the case, some of the other characters mentioned in the description, perhaps even the hispidity of the skin, may result from this circumstance. They must therefore be received with caution until larger specimens shall have been obtained.

* Since the above was printed, Mr. Waterhouse has been kind enough to show me in the museum of the Zoological Society the specimen which he believes to be the one procured by Mr. Cuming. Unless the characters are very much altered by age, it is decidedly distinct from the *A. hispidus* above described. The general form indeed is the same; but the skin is perfectly smooth, marked with vertical striæ; the granulated ridges on the head are less prominent, and the superciliary ridges without spines. The fin-ray formula is not quite as stated by Mr. Bennett, who appears, in his computation, to have mistaken the last dorsal spine for one of the soft rays of that fin, and also to have over-estimated the number of soft rays in the anal. The formula is really 18/12; A. 1/9, &c. I have no doubt of Mr. Cuming's fish being the true *A. Peruvianus;* whilst the one here characterized as new is probably the young of a nearly allied species. Mr. Cuming's specimen is six and a half inches long.

It may be advantageous to science to mention here, though not immediately connected with the present inquiry, that another species of *Agriopus* in the museum of the Zoological Society, which was seen by M. Valenciennes during his visit to this country, and referred by him in the "Histoire des Poissons" to the *A. verrucosus*, proves not to be that species, but the *A. spinifer* of Dr. Smith, recently described by him for the first time in his "Illustrations of the Zoology of South Africa."

FAMILY.—SCIÆNIDÆ.

OTOLITHUS GUATUCUPA. *Cuv. et Val.*

Otolithus guatucupa, *Cuv. et Val.* Hist. des Poiss. tom. v. p. 56. pl. 104.

FORM.—Elongated, with the back only very slightly elevated beneath the first dorsal: in advance of that fin the dorsal line is nearly straight, and continuous with the profile. Greatest depth contained exactly four times and a half in the entire length. Head long, contained three and a half times in the same. Lower jaw projecting considerably beyond the upper, and ascending to meet it. Two strongly developed curved canines at the extremity of the upper jaw; the rest of the teeth in this jaw consist of a single row of fine card, nearly equal throughout: in the lower jaw there are no canines, but one similar row of card, rather stronger than those above, and not equal, the smallest being in front, and those at the sides becoming gradually larger as they extend backwards. No vestige of scales on the lower jaw, lips or maxillary; but the suborbital is covered with bright silvery scales. Eye full, and moderately sized; its diameter one-fifth the length of the head; its distance from the end of the upper jaw equalling the diameter. Margin of the preopercle with a few indistinct striæ and obsolete denticulations. Opercle with two flat points not much developed. Lateral line very distinct, commencing at rather less than one-third of the depth, but curving gradually downwards to one-half; continued to nearly the extremity of the caudal; each scale marked with an elevated line, from which there proceed one or two small ramifications on each side.

First dorsal triangular, with the first spine very short, the fourth longest, the fifth and succeeding ones gradually decreasing, the last or tenth being shorter than the first. Second dorsal almost contiguous, its spine or first ray about equalling the first ray of the first dorsal: this fin is more than half as long again as the first, and the rays are nearly even. The anal commences further back than a point opposite the middle of the second dorsal; there are in reality two spines in this fin, but the first is so extremely minute as to be almost microscopic, and not seen, unless very carefully sought for; the second or principal spine is weak, and rather more than one-third the length of the soft rays. Caudal apparently square, but the rays being worn at the tips, its exact form cannot be determined. The second dorsal, as well as the anal and caudal, are partially covered with small scales, which, however, are not very obvious. Pectorals narrow and rather small, being scarcely more than half the length of the head. Ventrals placed a little further back, and rather shorter than the pectorals.

B. 7; D. 10—i/20; A. 1/8; C. 17; P. 16; V. 1/5.

Length 9 inc. 9 lines.

COLOUR.—"Silvery white, above iridescent with violet purple and blue."—D. Mr. Darwin has not noticed the dark transverse lines, which descend from the back obliquely forwards, as repre-

G

sented in the ' Histoire des Poissons,' and of which there are evident traces, though apparently much effaced by the action of the spirit.

Habitat, Maldonado Bay, Rio Plata.

This species, which Cuvier and Valenciennes consider as the *Guatucupa* of Margrave, was obtained by Mr. Darwin at Maldonado. M. D'Orbigny had previously taken it at Monte Video. The only respects in which Mr. Darwin's specimen differs from D'Orbigny's, is in its having two more rays in the soft dorsal, and a slightly longer anal spine, judging from the figure in the ' Histoire des Poissons ;' but I cannot imagine that they are distinct on these grounds only, so exactly do they agree in all their other characters.

CORVINA ADUSTA. *Agassiz.*

Corvina adusta, *Spix et Agass*. Pisces Brazil. p. 126. tab. 70.

FORM.—Greatest depth beneath the commencement of the first dorsal fin, and equalling one-fourth of the entire length. Back somewhat carinated, and moderately arched, forming one continuous curve with the profile, which falls with considerable obliquity. Ventral line nearly straight, and the abdomen much flattened in front of, and between the ventrals. Length of the head just equalling the depth of the body. Snout obtuse, with two small lobes at bottom, one on each side of the extremity, as in several other species of this genus. Mouth horizontal, at the bottom of the snout; when closed, the maxillary reaching a little beyond a vertical from the anterior margin of the orbit. Four pores beneath the symphysis ; and seven, in two rows, round the extremity of the snout; those in the lower row large. Jaws nearly equal ; the upper one perhaps a little the longest. Teeth forming a velutine band above and below ; those above with an outer row of somewhat longer and stronger ones. Eyes rather small ; their diameter about one-fifth the length of the head. Nostrils consisting of two round apertures in advance of the eye, the posterior one largest ; the anterior with a raised margin. Preopercle a little less than rectangular, with the angle at bottom somewhat rounded : the ascending margin rectilineal, sloping rather in advance of a vertical, and distinctly toothed, the teeth becoming smaller upwards : at the angle are two stronger teeth or spines, the uppermost directed backwards and a little downwards, the lowermost downwards and a little backwards ; between these two teeth there is an interval ; the basal margin of the preopercle is quite smooth. Opercle terminating in two flat inconspicuous points.

Snout, cheeks, and gill covers, covered with scales of very unequal sizes: those serving as a boundary between the cheek and the preopercle, also a row above each orbit, a few at the upper angle of the opercle, some on the suprascapular lamina, and a row extending thence upwards and forwards to the occiput, much smaller than the others. Scales on the body of moderate size, arranged in oblique rows; about fifty-five in a longitudinal line, and nineteen or twenty in a vertical. One taken from above the lateral line, and nearly in the middle of the length, is oblong, approaching to circular, its surface marked with a number of concentric, much crowded, curved lines, somewhat undulating behind, with a fan of about twelve deeper striæ converging to a point considerably in advance of the centre of the scale ; the free portion is also marked with several well-marked nearly parallel lines which terminate in denticles at the anterior margin. Those on the lateral line have the mucous tubes somewhat ramified, and are accompanied throughout its course by some minuter scales, similar to those on the head above pointed out. The lateral line is at one-third of the depth, till it arrives beneath the middle of the soft dorsal, where it falls to one-half.

First dorsal of a triangular form, separated from the soft portion by a deep notch ; the first spine very small and inconspicuous ; the second somewhat shorter than the third ; fourth longest, nearly equalling half the depth ; all the spines in this fin rather slender. The second dorsal commences with a spine somewhat longer than the last spine in the first dorsal, and not quite half the length of the first soft ray ; soft rays nearly even throughout, and not equalling the highest point of the first dorsal. Anal short and somewhat rounded, commencing beneath the middle of the second dorsal, and double the height of that fin ; its first spine very short and inconspicuous ; second long and moderately stout, but shorter than the first soft ray by one-third ; second soft ray the longest ; third and succeeding ones gradually decreasing. Pectorals narrow and pointed, shorter than the head ; first ray simple, the rest branched ; third, fourth and fifth longest. Ventrals attached a trifle backwarder than the pectorals, which they do not equal in length ; the spine much slenderer than that of the anal, and rather more than half the length of the first soft ray. Caudal squarish, but with the margin a little sinuous.

B. 7 ; D. 10—1/28 ; A. 2/8 ; C. 17 ; P. 17 ; V. 1/5.

Length 8 inches 6 lines.

COLOUR.—" Above inclining to coppery, with irregular transverse bars of brown ; beautifully iridescent with violet."—D. The bars alluded to by Mr. Darwin are some dark lines which, commencing at the upper part of the back, pass forwards and downwards in an oblique direction ; they bend more and more downwards as they advance, and disappear a little below the middle. The whole fish has a metallic gloss, particularly about the cheeks and gill-covers, and very visible even in its present state.

A second specimen, exactly similar to the above, is nearly twelve inches in length.

Habitat, Maldonado and Monte Video.

I entertain no doubt of this species being the *C. adusta* of Agassiz, figured in Spix's Fishes of Brazil. It is not described by Cuvier and Valenciennes, but belongs to their second section of the genus, characterized by the small spines on the ascending margin, and especially at the angle, of the preopercle. It seems to be particularly distinguished by the small scales on some parts of the head, and along the lateral line where they accompany the larger ones. These characters have not been overlooked by Agassiz. There are two specimens in the collection, the larger one taken at Monte Video, the smaller at Maldonado.

1. UMBRINA ARENATA. *Cuv. et Val.*

Umbrina arenata, *Cuv. et Val.* Hist. des Poiss. tom. v. p. 141.

FORM.—Rather elongated, with the back very little arched ; the greatest depth contained about five times and a quarter in the entire length. Length of the head about equal to the depth of the body. Profile falling very gradually, and nearly in a straight line, in front of the dorsal. Snout very much projecting ; the margin at bottom, above the upper jaw, divided into four lobes which are cut square at their extremities. Round the end of the snout, and immediately above the lobes, is a double row of pores, the lower ones large. Also four pores beneath the symphysis of the lower jaw. Barbule at the chin scarcely exceeding a line or a line and a half in length. A band of velutine teeth in each jaw, with an outer row in card ; these last moderately strong, sharp, and rather wide apart, not above fourteen or sixteen in the row. Preopercle very obsoletely denticulated. Opercle with two flat points not much developed.

First dorsal triangular ; the first spine very small ; the second, third and fourth elevated rather in a point, the third equalling two-thirds of the depth of the body or more. Second dorsal nearly twice the length of the first. Anal commencing opposite the sixth soft ray of that fin, short, and terminating considerably before it ; the anal spine weak, and very little more than half the length of the soft rays. Caudal with the posterior margin sinuous, the upper part being slightly crescent-shaped, the lower portion rounded, and broader than the upper. Pectorals a very little shorter than the head. Ventrals attached a little behind the pectorals, and not passing beyond them. In the axilla of the pectorals is a small triangular membranous lamina : there is also a narrow pointed one in the axilla of the ventrals covered with scales. The scales on the body are thin, rather small, somewhat rhomboidal, with their free margins ciliated, and with a fan of twelve striæ behind.

B. 7 ; D. 10—1/25 ; A. 1/8 ; C. 17 ; P. 21 ; V. 1/5.

Length 9 inches 6 lines.

COLOUR.—" Body mottled with silver and green : dorsal and caudal fins lead-colour."—D. *In spirits*, the colour appears dusky brown, with darker mottlings and silvery reflections ; paler beneath. The fins are dusky, but the basal half of the dorsal is darker than the upper. The pectorals are darker than the other fins, especially the inside ; on the left pectoral, the dark colour is restricted to three broad transverse fasciæ. There are also on the pectorals and anal, and on most of the scales on the body, small blackish dots, as mentioned in the ' Histoire des Poissons.'

A second specimen, smaller than the above, has the back rather more arched, the greatest depth being only five times in the length. The outer row of teeth in the upper jaw is not quite so conspicuous, the teeth being smaller and closer-set, and consequently more numerous. The soft dorsal and anal have fewer rays.

D. 10—1 22 ; A. 1/7 ; &c.

Length 7 inches 3 lines.

In all other respects similar to the specimen first described.

Habitat, Bahia Blanca, and Maldonado.

As Cuvier and Valenciennes have mentioned individuals of this species, which varied in the number of rays in the soft dorsal from twenty-two to twenty-four, I cannot but consider the two above described as specifically the same, though in the first these rays amount to as many as twenty-five. This, which is the larger specimen, was taken by Mr. Darwin at Bahia Blanca, where it is said to have been common. The other was obtained at Maldonado.

2. UMBRINA OPHICEPHALA. *Jen.*

U. elongata; rostro obtusissimo, tumido, haud ultrà fauces producto, margine inferiore quadrilobato, lobis intermediis rotundatis; fossulâ longitudinali inter nares, profundè exaratâ; poris quatuor infrà symphysin; dentibus velutinis, serie externâ in maxillâ superiore aculeiformi; preoperculo obsoletè denticulato; operculo mucronibus duobus parvis instructo; spinis dorsalibus tertiâ et quartâ longissimis, corporis altitudinem æquantibus; spinâ anali gracili, radiis articulatis dimidio breviori.

D. 12—1/22; A. 1/9; C. 17; P. 20; V. 1/5.

LONG. unc. 6. lin. 5.

FORM.—Very much elongated; the greatest depth just one-sixth of the entire length; the head one-fifth. Dorsal line nearly straight. Profile falling very slightly till it reaches the nostrils, when it suddenly becomes vertical. Snout in consequence short, and very blunt, and not projecting beyond the jaws; with a deep broad channel down the middle, extending from between the nostrils to near the mouth: on each side of this channel, the snout is very protuberant. The lower margin of the snout is divided into four lobes, the central pair of which are rounded: above each of the exterior lobes is one large pore, and an odd one in the middle. There are also four pores beneath the symphysis, and a short barbule, as in the last species. The eye has a diameter about one-fifth the length of the head, and is distant one diameter from the end of the snout. The nostrils, which are immediately in advance of the eye, consist of two round apertures, one before the other, the posterior one double the size of the anterior. Upper jaw a very little longer than the lower. A band of velutine teeth in each jaw; with an outer row above of moderately strong card, rather curving inwards and backwards, and closer-set than those of the *U. arenata*, amounting to twenty-eight or thirty in number: there are also some smaller card teeth behind this outer row passing insensibly into the velutine. Preopercle very obsoletely denticulated. Opercle with two flat points not very obvious.

First dorsal triangular, and moderately high in the point; third and fourth spines longest, about equalling the depth of the body; first spine very small: all the spines rather slender. Second dorsal about half as long again as the first, but the rays are too much broken to judge of their relative lengths. Anal spine very slender, and about half the length of the soft rays. The caudal is injured, but appears to have been of nearly the same form as in the *U. arenata.* The pectorals are about three-fourths the length of the head, but the ends of the rays are worn. The ventrals are of the same length as the pectorals in their present state: they are placed rather backwarder than in the *U. arenata,* being attached beneath the first third of the pectorals: there is a pointed scale in their axilla, of about the same relative size as in that species. The scales on the body are rather smaller, ciliated on their free edges, with a fan of eleven or twelve striæ behind. There are rows of small scales on the caudal, but none apparent on the other fins.

COLOUR.—Mr. Darwin did not notice the colours of this species in its recent state. *In spirits,* it appears of a nearly uniform dusky brown, but paler on the abdomen, with traces of silvery reflections about the head. The fins are dark, but the anal paler at the base than at the tips of the rays.

Habitat, Coquimbo, Chile.

This species may be at once distinguished from all those described in the 'Histoire des Poissons,' by its very elongated form. The head also has a peculiar character about it, and is not unlike that of some serpents. It appears to be the first species of this genus brought from the Pacific, the other foreign ones being all found either in the Indian seas, or on the Atlantic side of America. There are two specimens in the collection, exactly similar, and both obtained by Mr. Darwin at Coquimbo. They are, however, both in very bad condition; so much so, indeed, that I should have hesitated about describing them as new, had they not presented several obvious peculiarities.

GENUS—PRIONODES.* *Jen.*

Serrani formam quam maximè gerens. Pinna dorsalis unica, per totam longitudinem subæqualis. Membrana branchialis septem-radiata. Nec fovea, nec pori, infrà symphysin. Dentes maxillares velutini, serie externá cæteris fortiori, paucis, hic illic sparsis, subcaninis; palatini nulli. Preoperculum denticulatum. Operculum mucronibus tribus posticè armatum. Spina analis secunda fortis. Squamæ corporis ciliatæ; minutissimæ inter radios pinnarum verticalium, in seriebus dispositæ.

I am called upon either to establish this new genus among the *Sciænidæ,* or to break down one of the essential distinctions set by Cuvier between this family

---

* Serræ figuram habens. A πριων.

W Harvtre del.[t]

Fig. 1 *Pronodes fasciatus*
2. *Stegastes imbricatus*   Nat. Size

and the *Percidæ.* The form is so completely that of a *Serranus,*—which it resembles especially in its dorsal fin, head, maxillary teeth, form and armature of the pieces of the gill cover, and in the arrangement of the scales on the body,— that at first sight no one would hesitate to refer it to that group; but *the vomer and palatines are without teeth.** In this respect, indeed, I consider it an important discovery; as it affords another striking instance of the uncertainty of this character, in cases in which others, which have been generally made subordinate to it, remain constant. It is probable that the time will come, when it will be found necessary to revise some portion of the *Percidæ* and *Sciænidæ* with reference to a more correct valuation of this character. For the present, however, I refrain from interfering with the Cuvierian arrangement; and the only alternative is to consider this as a new form among the *Sciænidæ,* where it must be placed along with those genera possessing one dorsal fin, and having seven rays in the branchiostegous membrane. Such are *Hæmulon, Pristipoma,* and *Diagramma;* from all which, however, it is at once distinguished by the absence of pores at the symphysis and on the lower jaw, and by the much more developed spines on the opercle, and from *Pristipoma* by its having, further, scales on the vertical fins. On the whole, it seems to approach nearest to *Hæmulon;* but the crown and snout are more free from scales than in that genus, and the scales on the body are not set in oblique rows, as is the case in so many of the true *Sciænidæ.* The head also has no cavernous appearance about it. This new form is from the Galapagos Archipelago.

PRIONODES FASCIATUS. *Jen.*

PLATE IX. Fig. 1.

P. pallidè flavescenti-fuscus, fasciis transversis plurimis suprà rubescenti-nigris infrà miniatis; pinnis verticalibus maculis parvis ocellatis: vertice, rostro, et maxillis, nudis; preoperculo margine adscendenti denticulato, basali lævi; operculo mucrone intermedio forti; spinis dorsalibus ad apices laciniis investitis; pinnâ caudali subæquali.

B. 7 ; D. 10/12 ; A. 3/7 ; C. 17 ; P. 18 ; V. 1/5.

LONG. unc. 7. lin. 3.

FORM.—Oval, compressed; the back not much arched, forming one continuous curve with the profile, which falls gently from the nape; ventral line less convex than the dorsal. Greatest

---

* With the exception of a small rough oblong spot, near the posterior extremity of the left palatine.

depth equalling one-fourth of the entire length; head about one-third; thickness rather less than two-thirds of the depth. Mouth rather wide, with the lower jaw longest. The maxillary dilates at its posterior extremity; it reaches to nearly beneath the middle of the orbit, and does not retire beneath the suborbital. In each jaw a band of velutine teeth; above there is an outer row of longer ones in card, and one or two in front on each side still longer resembling small canines; in the lower jaw there are also a few longer ones, of the same character as these last, interspersed at intervals. Tongue free at its extremity, and, as well as the vomer and palatines, without teeth. Eyes rather high in the cheek; their diameter about one-sixth that of the head. The nostrils consist of two small round orifices a little in advance of the eyes, the anterior one covered by a membranous flap. Margin of the suborbital entire. Preopercle finely denticulated on its ascending margin, which is vertical and slightly convex; but the denticulations almost disappear at the angle, and are not visible at all on the basal margin. Opercle triangular, with three flat spines, the middle one longest, beyond which the membrane projects in the form of an angular process to the distance of three lines. Small scales on the cheeks and preopercle; but none on the crown, snout, first suborbital, maxillary, or lower jaw; scales on the opercle larger, equalling those of the body in size. Gill-opening large, with the branchial membrane deeply notched in the middle.

Lateral line following the curvature of the back at one-fourth of the depth. Scales on the body moderately large: one taken from the middle of the side above the lateral line is of a somewhat oblong form, with the free edge rounded and finely ciliated; the basal portion with fourteen slightly converging striæ, which form at the hinder margin as many, but not very distinct, crenations.

The dorsal commences above the terminating lobe of the opercle, and reaches to within a short space of the caudal : height of the spinous portion, which, with the exception of the first two spines, is nearly even throughout, about one-third of the depth; soft portion rather higher, with the last two rays but one longest, and forming a point backwards; all the soft rays branched. Anal commencing in a line with the soft portion of the dorsal, and terminating a little before that fin; three spines, the second one-third longer than the first, and a little longer than the third, and much the strongest of all; the soft portion of the anal is similar to that of the dorsal, and terminates in like manner in a point behind. Space between the anal and caudal a little less than one-sixth of the entire length. The caudal appears to have been nearly even, or perhaps slightly rounded, but the rays are worn. Rows of very minute scales, not very obvious, between the rays of all the vertical fins. Pectorals slightly rounded; more than half the length of the head; all the rays with the exception of the first two and the last, branched. Ventrals attached beneath, or perhaps a very little in advance of the pectorals; pointed, with the second soft ray longest. No lengthened scale or process of any kind in the axilla of either ventrals or pectorals; neither are the former fastened to the abdomen by a membrane half their own length, as is the case in many of the *Serrani*.

Colour.—" Pale yellowish brown, with numerous transverse bars, of which the upper part is reddish black, the lower vermilion red ; gill-covers, head, and fins, tinted with the same."—D. Mr. Darwin has not noticed some small round black spots surrounded by a white border, and having an ocellated appearance, which are very evident on the upper half of the soft portion of the dorsal : there is a faint indication of similar spots on the anal and caudal.

Habitat, Chatham Island, Galapagos Archipelago.

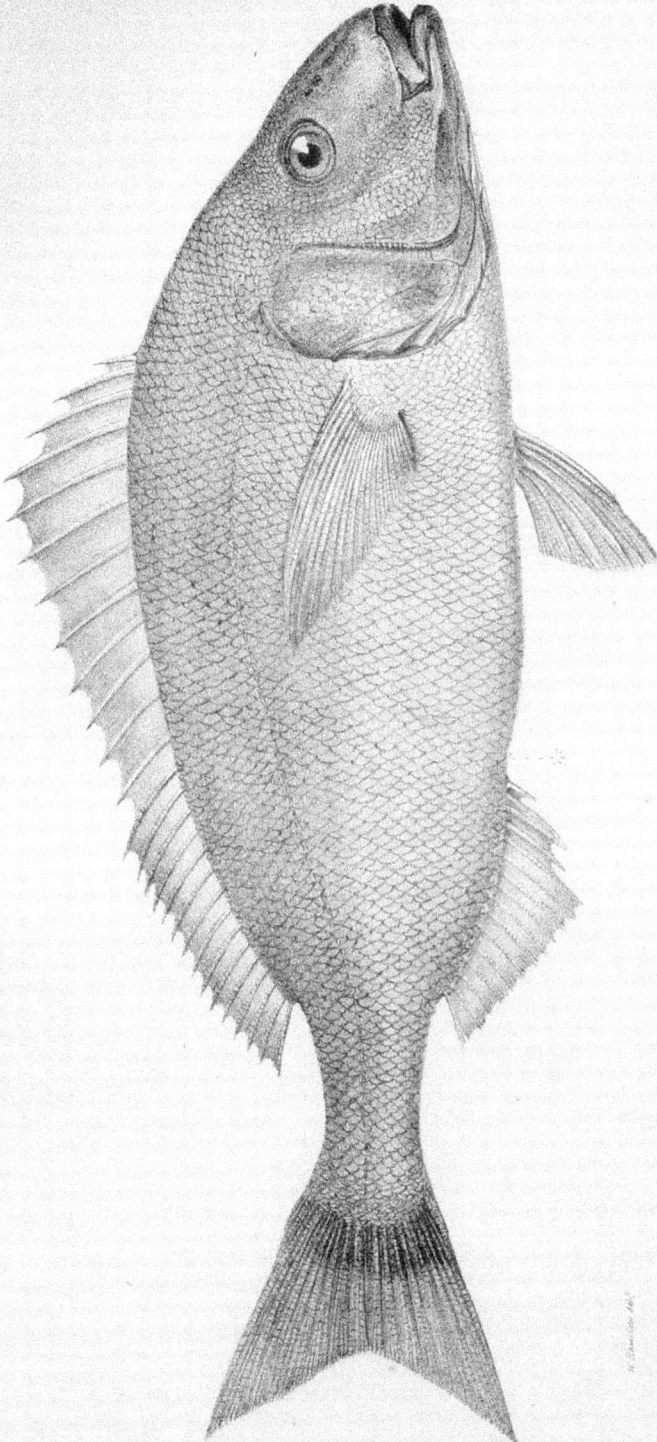

M. Dubourjal del.

Pristipoma cantharinum Nat size

Mr. Darwin obtained one specimen only of this new genus at Chatham Island in the Galapagos Archipelago. It is probably not full-sized.

PRISTIPOMA CANTHARINUM. *Jen.*

PLATE X.

P. cæruleo-argenteum, operculo nigro-marginato: pinnâ dorsali subæquali, spinis ultimis radiis articulatis paulo brevioribus; anali spinâ secundâ forti, longitudinaliter striatâ, radiis articulatis duodecim : preoperculo rectangulato, margine adscendenti, leviter denticulato, basali integro : vertice, buccis, et ossibus opercularibus, squamatis; rostro ultrà nares, suborbitalibus, et maxillis, nudis; squamis corporis ciliatis: pinna caudali furcatâ.

B. 7; D. 12/15; A. 3/12; C. 17, et 4 breviores; P. 20; V. 1/5.

LONG. unc. 10. lin. 11.

FORM.—Form oblong-oval, much resembling that of the *Cantharus griseus.* Body compressed, with the dorsal line slightly curved ; the profile descending from the nape more obliquely, and in a very regular manner. Greatest depth beneath the commencement of the first dorsal, contained not quite three times and three quarters in the entire length : head rather less than one-fourth of the length. Mouth protractile, but not wide, the commissure not extending to a vertical from the anterior angle of the eye; when closed, the maxillary retires beneath the suborbital, and only just the extremity remains visible. Jaws equal ; in each a narrow band of velutine teeth, the outer row somewhat longer than the others, particularly above, where they approach to card. Tongue, palatines, and vomer smooth. Eyes moderate ; their diameter rather less than one-fifth of the length of the head ; rather nearer the extremity of the snout than the posterior margin of the opercle ; the distance between them equalling twice their diameter. Two small pores and a fossule beneath the symphysis of the lower jaw, the latter very distinct. Preopercle rectangular, the angle somewhat rounded ; the ascending margin nearly straight and finely denticulated, but the denticulations hardly continued to the angle, and not appearing at all on the basal margin. Opercle with two small flat points, but very indistinct and almost lost in the membrane. Suborbitals large, with their lower margins entire. Crown, cheeks, and pieces of the gill-cover, covered with small scales; but not the snout in advance of the nostrils and eyes, nor suborbitals, nor lower jaw. Suprascapulars marked by a large scale, the margin of which is nearly entire.

Lateral line following the curvature of the back at one-third of the depth ; each scale marked with an elevated line without ramifications. A scale taken from above the lateral line is of a somewhat rhomboidal form ; the free portion very finely striated, with the margin finely

H

ciliated; the concealed portion with eight or nine deeper and more distinct striæ, not meeting in the centre to form a fan, and with the basal margin crenated. The scales on the cheeks and opercle are smaller than those on the body, and almost smooth.

The dorsal fin commences in a line with the posterior margin of the opercle, and extends nearly the whole length of the back, rising from a groove as in the *Sparidæ:* its height on the whole tolerably uniform throughout: spinous portion occupying more than half the fin; the anterior spines gradually increasing in length to the fourth,* which equals rather more than one-third of the depth; the succeeding ones nearly even, very gradually decreasing to the last, which is about two-thirds the length of the fourth; all the spines moderately stout: soft portion of the fin even, and rather higher than the last spine. Anal commencing in a line with the third soft ray of the dorsal, and terminating opposite to that fin: the first spine short, but strong; second and third spines equal in length, being about two-thirds the length of the soft rays, but the second much stouter than the third; the second spine is also distinguished from the others by having its surface longitudinally striated: soft rays nearly even, and resembling those of the dorsal. Caudal forked, with the upper lobe a trifle longer than the lower; the basal half covered with minute scales. Pectorals narrow and pointed, about two-thirds the length of the head, with a small fold of loose skin in their axillæ. Ventrals placed a little further back than the pectorals, and somewhat shorter; a long pointed scale in their axillæ, nearly one-third their length.

COLOUR.—" Bluish silvery."—D.—The colour, as it appears in spirit, is nearly uniform bluish gray, and very similar to that of the *Cantharus griseus.* The gill-cover has a dusky edging posteriorly.

Second specimen.—Smaller than the above, and not quite so deep in the body; the greatest depth contained a trifle more than four times in the entire length; the nape in consequence less elevated, and the profile less oblique. Eyes relatively a little larger, their diameter rather more than one-fifth the length of the head. Preopercle with the posterior margin not so rectilineal, approaching to concave; the angle at bottom projecting in consequence a little backwards; the denticulations not quite so distinct and regular. One ray more in the soft dorsal.

D. 12/16; A. 3/12; C. 17, &c.; P. 19; V. 1/5.

LONG. unc. 9. lin. 2.

COLOUR.—" Silvery; above, shaded with brown and iridescent with blue; fins and iris sometimes edged with blackish brown. Flap of the gill-cover edged with black."—D.

Habitat, Galapagos Archipelago.

This species, which is undoubtedly new, may be known from most of those described by Cuvier and Valenciennes by its greater number of soft rays in the anal fin. The only ones which equal it in this respect are the *P. Conceptionis* and

---

* The third spine is broken, and may have been as long as the fourth.

the *P. fasciatum*; from the former of which it may be distinguished by its greater depth and nearly even dorsal, from the latter by its plain colour free from all conspicuous bands and markings. The dorsal notch is scarcely observable, the eleventh and twelfth spines being nearly equal, and but little shorter than the first soft ray. Its analogy to the genus *Cantharus* among the *Sparidæ*, which it resembles as well in colour as in general form, is very striking. There are two specimens in the collection; the one described first above having been taken at Chatham Island, the other at Charles Island, in the Galapagos Archipelago.

1. LATILUS JUGULARIS. *Val.*

Latilus jugularis, *Cuc. et Val.* Hist. des Poiss. tom. ix. p. 369. pl. 279.

FORM.—Elongated, with the dorsal line slightly curved, the ventral nearly straight. Greatest depth contained five times and one-third in the entire length. Head, which much exceeds the depth, four times in the same. Profile very convex above the eyes, whence it falls obliquely to the lips. Snout thick and rounded, resembling that of the *Red Mullet*: mouth protractile, horizontal, placed at the bottom of the snout, the commissure just reaching to a vertical from the anterior part of the orbit. Jaws equal or very nearly so; the lower one perhaps a very little the longest. Maxillary not widening at its posterior extremity. A band of velutine teeth in each jaw, narrowing at the sides as it extends backwards; with an outer row of longer and stronger ones: in the lower jaw, the velutine band does not extend beyond the middle of the sides, the carding teeth being all that are visible. Tongue and palate smooth. Eyes high in the cheeks; large, and of an oval form; their vertical diameter three-fourths of their longitudinal; this last equalling one-fourth the length of the head. Nostrils consisting of two round apertures, the posterior one largest, the anterior covered by a membranous flap. Preopercle with the denticles far apart, and not very obvious, unless the skin be dissected off; the ascending margin rectilineal and vertical; the angle rounded. Bony part of the opercle terminating in a flat point, above which are two other smaller points not so well developed; all the points concealed in the membrane, and scarcely visible from without: beneath the principal point, the membrane is prolonged backwards in the form of a broad flattened bristly point three lines in length. Crown, gill-covers and cheeks, scaly, but not the jaws; snout scaly, except very near the lips. Gill-opening large.

Lateral line at first at one-third of the depth, but falling gradually to one-half. Scales rather small; one taken from immediately above the lateral line of an oblong form, the length being twice the breadth, with its free margin finely ciliated, crenated behind with a fan of nine striæ; on scales taken from other parts the number of striæ in the fan are more numerous.

One long dorsal fin of nearly uniform height throughout, equalling about half the depth; only four slender spines, gradually increasing in length from the first which is very short; the fourth about three-fourths the length of the first soft ray; soft rays increasing likewise very gradually to the fourth, which with the next five or six are highest; the membrane of the fin very delicate; all the soft rays branched. Vent in a vertical line with the ninth soft ray of the dorsal. Anal commencing immediately behind it, and answering to that portion of the dorsal to which it is opposite, terminating at the same distance from the caudal; only two slender spines, the first very short; the first soft ray simple, the rest branched. Space between these two fins and the caudal barely one-eighth of the entire length. Caudal nearly even. Pectorals moderately long and narrow, equalling nearly the length of the head; rays branched; fourth, fifth, and sixth longest. Ventrals a little in advance of the pectorals, nearly equalling them in length; of a pointed form, with the third and fourth soft rays longest. In the axilla of the pectorals a vertical scaly membranaceous lamina.

B. 6; D. 4/28; A. 2/22; C. 17; P. 20; V. 1/5.

Length 11 inc. 5 lines.

COLOUR.—(*In spirits.*) Dusky olive on the back and upper part of the sides, yellowish (probably silvery in the recent state) beneath, with faint indications of five or six dark transverse bands, similar to those in the *common perch.* Inside of the ventrals blue.

Second specimen.—Smaller than the above, measuring six inches and a half in length, but differing from it in no respect, as regards form, excepting in having the profile not so oblique, and the snout in consequence not so obtuse; the jaws also are exactly equal. Fin-ray formula the same.

COLOUR.—" Beneath brilliant white; head and back clouded with purplish and carmine red; longitudinal and transverse irregular bands of the same."—D. The bands in this specimen amount to eight in number, and are much more conspicuous than in the larger one above described.

Habitat, Valparaiso, Chile.

The smaller of the two specimens above described was taken by Mr. Darwin at Valparaiso. The number attached to the larger one has been lost, but it was probably taken at the same place, where it had been previously discovered both by M. D'Orbigny and M. Gay. The specimen described by Valenciennes has one soft ray more in the dorsal, and one less in the anal, than either of the above; but in all other respects they tally exactly. As observed in the " Histoire des Poissons," this species has many points of resemblance to *Percis* and *Pinguipes.*

2. LATILUS PRINCEPS. *Jen.*

PLATE XI.

L. elongatus; corporis altitudine capitis longitudinem æquanti; dentibus velutinis, serie externâ fortiori, aculeiformi; præoperculo margine adscendenti recto, leviter

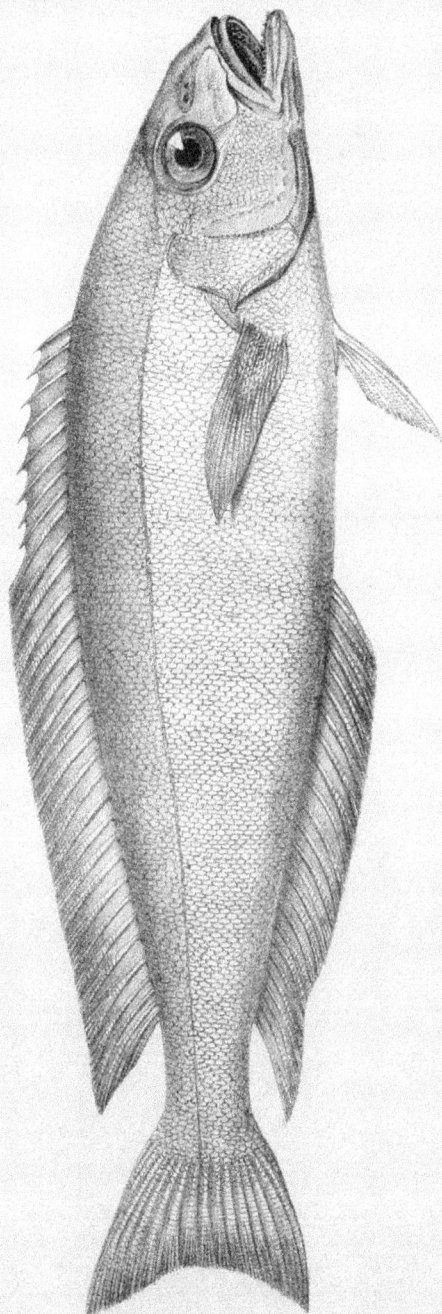

Latilus princeps, ½ nat. size.

W. Bowbear del.t

denticulato, basali lævi; operculo mucrone unico; rostro, ossibus suborbitalibus, maxillis, limbo preoperculi, et interoperculo, nudis; buccis et cranio squamatis, squamis in vertice spatium angulatum inter oculos occupantibus; pinnis dorsali analique prælongis; spinis analibus parvis, gracilibus, primâ minutissimâ; ventralibus accuratè thoracicis; cauduli emarginatâ.

B. 5?; D. 8/26; A. 2/26; C. 15, &c.; P. 18 vel 19; V. 1/5.

LONG. unc. 20. lin. 6.

FORM.—Elongated; the greatest depth equalling the length of the head, and each contained rather more than four times and a half in the entire length. Snout short and rather obtuse, the profile bending downwards in a curve before the eyes. Mouth nearly horizontal, at the bottom of the snout; when closed, the maxillary, which is not widened at its posterior extremity, and which is very similar in form to that of the last species, reaches nearly, but not quite, to a vertical from the anterior part of the orbit. Lower margin of the suborbital entire. Teeth forming a velutine band in each jaw, widest in front, with a row of stronger ones externally: none on the tongue, vomer, or palatines. Eyes large, and high in the cheeks; their diameter one-fifth the length of the head. Preopercle with the angle at bottom rounded; the ascending margin straight, and nearly but not quite vertical, forming with the basal rather more than a right angle; the former finely denticulated, but not the latter. Opercle terminating in one flat point, not projecting beyond the membrane. The branchiostegous rays appear to be but five in number, but, the skin being dry, there may possibly be a sixth overlooked. Cranium, cheeks, and opercle scaly; but not the snout or jaws, or limb of the preopercle, or interopercle: the scales on the crown are separated from the naked skin of the snout by a well-defined line, which forms an advancing angle between the eyes.

Lateral line straight, and continued to the base of the caudal; its course parallel to the back at between one-fourth and one-third of the depth. Scales on the body rather small, oblong, longer than broad, with their free extremities dotted and finely ciliated; the concealed portion striated finely at the sides, and more deeply at the base; but all the central portion, including an oblong area of the same form as the entire scale, without striæ, being only very minutely roughened or punctured.

One long dorsal, low, and of nearly uniform height throughout, commencing about in a line with the insertion of the pectorals, and reaching very nearly to the caudal: eight spines, rather slender, and very gradually increasing in length, the last being just twice the length of the first and equalling the distance from the base of the fin to the lateral line: the soft rays which follow are nearly even with the last of the spinous till the twenty-fourth, which is slightly prolonged in a point, and which is followed by two others shorter than the rest; the ends of the rays are rather worn, but they appear to have been all branched. Anal also long, commencing at about the middle of the entire length, or in a line with the sixth soft ray of the dorsal, and terminating opposite to that fin, to the last half of which, or rather more than half, it exactly answers; only two spines, which are so slender and minute, especially the first, and so closely united to each other as well as to the first soft ray, as to be scarcely obvious except upon dissection ; all the soft rays, except the first, branched. Space between the anal and caudal not a tenth part of the whole length. Caudal slightly notched, or hollowed out, with rows of scales between the rays. Pectorals pointed, about three-fourths the length of the head, with the seventh and eighth rays longest ; rows of scales at the base between the rays : in their axillæ a somewhat projecting vertical scale or lamina, as in the last species. Ventrals immediately beneath the pectorals, also pointed, but shorter.

Colour.—" Above, and the fins, obscure greenish ; sides obscure coppery, passing on the belly into salmon-colour. Pectorals edged with dull blue. Iris yellowish brown : pupil black-blue."—D.—The skin has dried to a nearly uniform brown.

Habitat, Chatham Island, Galapagos Archipelago.

I feel but little hesitation in referring this species, which is one of the many new ones obtained by Mr. Darwin in the Galapagos Archipelago, to the genus *Latilus*. The absence of vomerine and palatine teeth requires it to be placed, according to Cuvier's views, among the *Sciænidæ;* in which family, there is no other group besides *Latilus,* to which it makes any approach. It agrees with that genus in its general form, and in many of its particularities ; it has the same form of snout, mouth, maxillary, and dentition ; the same scaly lamina in the axilla of the pectorals ; the same long undivided dorsal and anal fins, with only two very small anal spines, so closely united to the first soft ray as to be easily overlooked. But it may be at once distinguished from the *L. argentatus* and the *L. doliatus,* the only two species described by Cuvier and Valenciennes in the body of their work, by its much more numerous soft rays in the dorsal and anal fins. From the *L. jugularis* last described, which resembles it in this respect, it differs in its thoracic ventrals, shorter head, naked snout and suborbital, and notched caudal : the profile also falls less obliquely. There is only one specimen in the collection, a dried skin and rather injured.

HELIASES CRUSMA. *Val.*

Heliases Crusma, *Cuv. et Val.* Hist. des Poiss. tom. ix. p. 377.

Form.—Oval, very much compressed. Back considerably elevated, particularly at the nape, whence the profile descends very obliquely, and, with the exception of a slight concavity before the eyes, in nearly a straight line. Greatest depth at the commencement of the dorsal, equalling nearly half the entire length, caudal excluded. Head contained four and a half times in the same. Snout short : mouth small, a little protractile : lower jaw rather the longest. A narrow band of velutine teeth in each jaw, with the outer row in fine card ; these last longest and strongest in front.   Eyes large; their diameter nearly one-third the length of the head.   Sub-orbitals forming a narrow curved band beneath the eyes, and covered by a row of scales. Nostrils with only a single, small, round aperture.   Preopercle with the ascending margin vertical, not quite rectilineal, inclining slightly inwards towards the angle, which is rounded. Opercle, taken together with the subopercle, very regularly curved, the margin describing nearly a semicircle, with one flat point to terminate the osseous portion ; its height double its length.

The whole of this fish, including every part of the head, except the lips and maxillary, is covered with scales, which extend on to the vertical fins as in *Glyphisodon* : those on the fins and upper part of the head and snout are very small, but those on the gill-covers and body very large : about twenty-six or twenty-seven in a longitudinal line from the gill to the base of the caudal, and fourteen or fifteen in a vertical line: one taken from about the middle of the side is oblong, the breadth exceeding the length, with the anterior margin rounded, and the free portion finely dotted and very minutely ciliated, the concealed portion cut square, with a fan of eight or ten striæ not meeting at the centre, and terminating at the basal margin in as many crenations.   The lateral line commences at one-fourth of the depth, but, from the fall of the dorsal line posteriorly, the distance between these two lines diminishes as the former advances : the lateral line terminates beneath the soft portion of the dorsal fin altogether.

Fins almost exactly similar to those of the *Glyphisodon saxatilis* and *Heliases insolatus*, as described and figured in the " Histoire des Poissons."   The fourth and fifth spines in the dorsal longest, equalling one-fourth of the depth ; of the soft rays the third, fourth, and fifth are longest.   First anal spine only one-third the length of the second, which is itself rather shorter than the soft rays ; and these last appear longer than in the *H. insolatus*.   Caudal more forked than crescent-shaped, the depth of the fork equalling nearly half the length of the fin, which is itself one-fourth the entire length of the fish.   Axillary scales of the pectorals and ventrals as in *H. insolatus*.

B. 6 ; D. 13/12 ; A. 2/12 ; C. 15, & 4 short ; P. 21 ; V. 1/5.

Length 8 inches.

Colour.—" Above lead-colour, beneath paler."—D.   In spirits, it appears of a deep brownish olive on the back and upper part of the sides, passing into dull golden yellow on the lower part of the sides and abdomen, where, however, the scales are still faintly edged with the former colour.   Fins dark.

Habitat, Valparaiso, Chile.

This species, as M. Valenciennes observes, is so extremely similar to the *H. insolatus*, that at first sight, it would hardly be distinguished from it.   The only differences appear to consist in the form of the caudal, which is forked, not crescent-shaped as in the species just mentioned, and in the greater length of the soft rays of the anal. In the figure of *H. insolatus* in the " Histoire des Poissons," these rays are represented of the same length as the second spine, whereas in the present species they rather exceed it, giving a greater depth to the entire fin. The teeth also would seem to be more developed in the *H. Crusma*, especially those in front, which are longer than the others. According to Valenciennes, the outer row hardly exceeds the inner ones in the *H. insolatus*. The geographical position of the two species is however widely different. The *H. insolatus* is a native of the Caribbean Seas ; whereas the *H. Crusma* has only been obtained on the coast of Chile and off the island of Juan Fernandez. M. Gay first obtained it at Valparaiso, where also Mr. Darwin's specimen was procured ; in whose notes it is stated, that it gets to a much larger size than the one here described.

FAMILY—SPARIDÆ.

CHRYSOPHRYS TAURINA. *Jen.*

PLATE XII.

C. albida, quatuor fasciis interruptis nigro-fuscis ; pinnis dorsali, caudali, et ventralibus, clarè cæruleo-marginatis: dentibus anticis conicis, in maxillâ superiore octo, in inferiore decem minoribus ; molaribus suprà seriebus tribus, intermediâ minori, infrà duábus dispositis ; preoperculo et operculo, utroque quatuor squamarum seriebus tecto ; limbo preoperculi nudo.

D. 12 12 ; A. 3/10 ; C. 17, &c. ; P. 15 ; V. 1 5.

LONG. unc. 14.

FORM.—General form not very dissimilar to that of the *C. Aurata.* Greatest depth contained about three times and a half in the entire length. Depth and length of the head equal, each about one-fourth of the entire length. Profile very oblique. Eyes high, and moderately large, distant two diameters from the end of the snout. Preopercle with the angle at bottom very much in advance, giving an obliquity to the ascending margin ; the limb not very broad, and naked ; in front of the limb are about four rows of scales smaller than those on the body : the same number of rows of scales on the opercle. Jaws equal, with eight conical incisors in front of the upper one, and ten in front of the lower ;[*] those above longer than those below, and more

[*] There are actually nine, but one appears to have been lost.

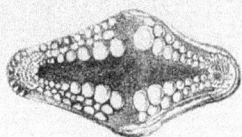

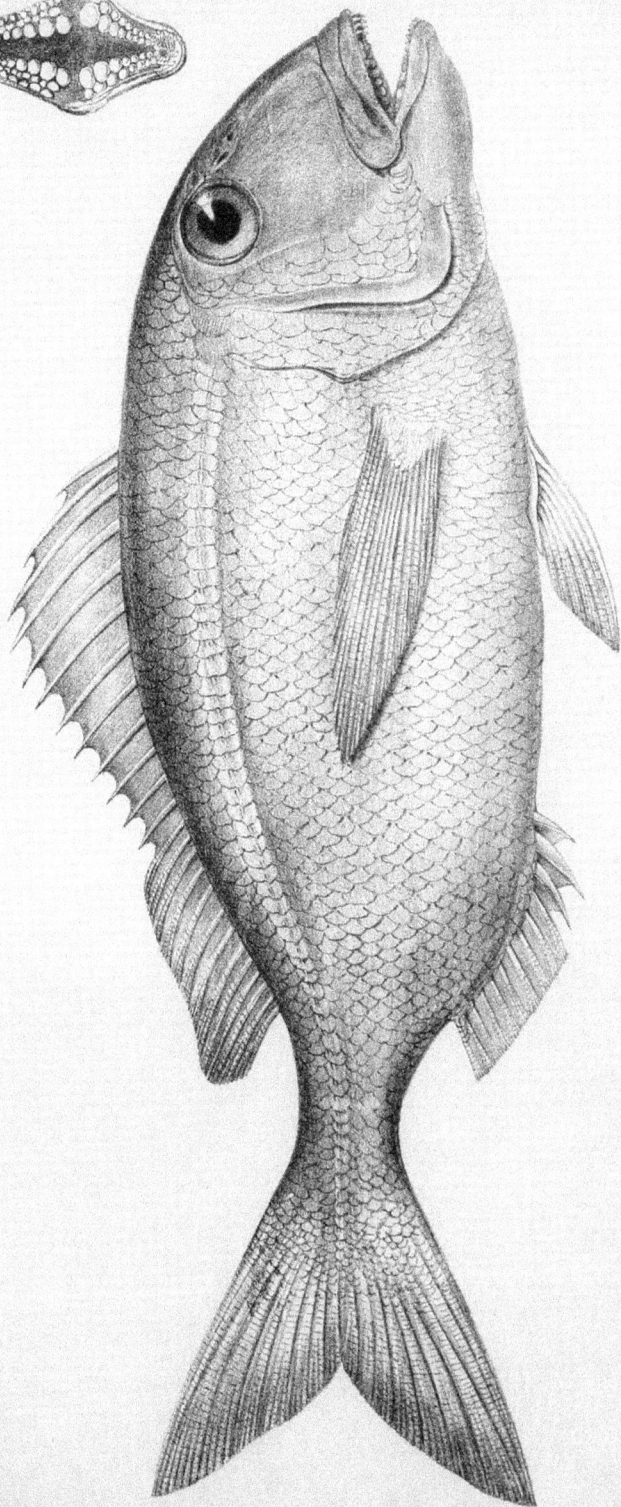

Chrysophrys taurina 3/4 Nat. Size

D. Hawkins del.

regularly and closely set : behind the incisors above and below is a patch of fine card : then follow the molars, which are in three very regular rows above and two below ; of the three rows above the inner and outer ones are much the strongest, containing each about eight teeth ; those in the outer row are slightly pointed, and not very unequal in size, but the inner series enlarge very rapidly as they extend backwards, the last two or three being of considerable size ; all round or nearly so, there being no large oval one at the back, as in the *C. Aurata* and some other species ; the middle series above consists of teeth much smaller than the others, and more numerous : the two rows below are not very dissimilar to the inner and outer rows above. Suborbital broad, and naked, covering a large portion of the cheek.

Scales on the body of a moderate size ; too much injured and displaced in this specimen to admit of the exact number being counted in a longitudinal row; those on the lateral line, however, are all perfect and present to within five rays of the end of the dorsal, and up to that point they amount to thirty-one. The fins, so far as can be judged from their present state, are on the whole very similar to those of the other species ; but the dorsal and anal spines, especially the second anal spine, appear rather stronger than those of the *C. Aurata.* Pectorals long and narrow, contained about three times and three quarters in the entire length.

Colour.—" White, with four dark brown much interrupted bands, giving a mottled appearance ; head coloured with the same ; top of the head, ridge of the back, edges of the dorsal, caudal and ventral fins, tinted with fine azure blue."—D.

Habitat, Chatham Island, Galapagos Archipelago.

Mr. Darwin's collection contains a single specimen of a species of *Chrysophrys* from the Galapagos Archipelago, not in a sufficiently good state of preservation to admit of a very detailed description being given of it, but, nevertheless, evidently distinct from any that I can find recorded by authors. It appears to belong to Cuvier's second section of this genus characterized by the absence of any large *oval* molar behind the others, though the last two or three in the inner series above are of considerable size. It differs, however, from all those described in the " Histoire des Poissons," in having the conical incisors more numerous, and but three rows of molars in the upper jaw. The specimen also is of sufficient size to lead to the belief, that it would not have acquired any additional ones by further growth. The *C. aculeata* resembles it, indeed, in this last character, but independently of other differences, this species is said to have a reclined spine before the dorsal fin which is not present in the one here described.

Out of twenty-two species of this genus described in the " Histoire des Poissons," only one is from the Pacific Ocean, whence the present species was brought. The greater number are from the Atlantic and Indian Oceans.

Family—MÆNIDÆ.

1. Gerres Gula. *Cuv. et Val.?*

Gerres Gula, *Cuv. et Val.* Hist. des Poiss. tom. vi. p. 349.

Form.—Greatest depth one-fourth of the entire length. Back but little elevated. Space between the eyes flat, with a fovea in the middle, which is prolonged in a channel nearly to the extremity of the snout. Length of the head exceeding its depth by one-fourth, and contained about three times and three quarters in the entire length. So much of the maxillary as is visible is of an oval form, its length being twice its breadth at its posterior extremity. Suborbital with the lower margin very indistinctly notched, and not denticulated. Eyes very large, their diameter contained twice and three quarters in the length of the head. The two orifices of the nostrils of nearly equal size. No denticulations on any of the pieces of the gill-cover. A narrow band of very minute velutine teeth in each jaw, those above hardly visible to the eye, but sensible to the touch: none on the vomer, palatines, or tongue.

Dorsal with the first spine extremely short; the second has a small piece broken off at the tip, but appears to have been about the same length as the third, which last equals two-thirds of the depth of the body; the fourth and fifth are a little shorter than the third; the succeeding ones gradually decreasing, as in the other species of this genus: all the spines are moderately slender, the anterior ones slightly arcuate, with scarcely any appreciable difference in the degree of stoutness in the first four. Anal with the first spine extremely short; the second obviously stouter than any of the dorsal spines, but much shorter, being only half the length of the second dorsal spine, or one-third the depth of the body; the third spine is a trifle longer than the second, but much slenderer. Caudal deeply forked; the lobes worn at the tips in this specimen, but their length, when perfect, probably about one-fourth, or somewhat less, of the entire length of the fish. Pectorals narrow and pointed, a little shorter than the head, and contained four and a half times in the entire length; fifth ray longest. Ventrals a little behind the pectorals, and not more than two-thirds their length, or scarcely so much; the spine a little shorter than the soft rays, and of about the same degree of stoutness as the dorsal spines. Elongated scale in the axillæ of the ventrals about three-fourths the length of the spine, of a narrow lanceolate form, ending in a very fine point.

D. 9.10; A. 3.7; C. 17, &c.; P. 14; V. 1.5.

Length 3 inc. 6 lines.

Colour.—Not noticed in the recent state. In spirits, it appears of a uniform silvery, with the back and upper part of the sides inclining to dusky olive: no bands or any particular markings: fins pale.

Habitat, Rio de Janeiro.

The species of this genus are numerous, and extremely similar to each other. Many of them appear to rest on characters taken simply from the relative lengths and degrees of stoutness of the dorsal and anal spines. This renders it extremely difficult to identify single specimens. Perhaps I am wrong in referring the one described above to the *G. Gula* of Cuvier and Valenciennes; but it makes so near an approach to that species, that I hardly dare characterize it as distinct. It cannot be the *G. Aprion* of those authors, which is closely allied to the *G. Gula*, and is found on the same coasts, since its teeth are so very much finer: the caudal also is not scaled. It is small, but Cuvier and Valenciennes state that none of their specimens of the *G. Gula* exceed five inches. Mr. Darwin took it in a salt-water lake, Lagoa de Boacica, at Rio de Janeiro.

2. GERRES OYENA. *Cuv. et Val.?*

Gerres Oyena, *Cuv. et Val.* Hist. des Poiss. tom. vi. p. 355.
Smaris Oyena, *Rüppell,* Atlas zu der Reise im Nörd. Afr. Zoologie; p. 11. tab. 3. fig. 2.

FORM.—Greatest depth contained rather more than three and a half times in the entire length: the dorsal curve very regular. Profile above the eyes a little concave. Length of the head exceeding its depth. Maxillary as in the species last described. Suborbital with its lower margin distinctly but not very deeply notched; not denticulated. Diameter of the eye less than one-third the length of the head. Posterior orifice of the nostrils twice the size of the anterior one. No denticulations on any part of the head or gill-cover. A narrow band of velutine teeth in each jaw, of about the same length and degree of fineness above and below; but none on the palate or tongue.

The dorsal commences in an exact vertical line with the insertion of the ventrals: the anterior spines are a little arcuate; the first, as in the other species of this genus, is extremely short; the second and third in this specimen are broken at their extremities so that their exact length cannot be ascertained, but the portion of the second remaining (and of this spine apparently only a very small piece is gone) nearly equals half the depth of the body; length of the fourth spine which is perfect not quite equalling two-fifths of the depth; fifth, sixth, and seventh spines gradually decreasing; eighth and ninth scarcely shorter than the seventh: the second spine is much compressed, and though obviously stronger than any of those which follow, not nearly so stout as in many other species; its breadth is not more than one-twelfth of its length. Anal commencing in a line with the fourth soft ray of the dorsal; the second spine compressed similarly to the second dorsal spine, and of about the same degree of stoutness, but its length one-third less, being just equal to one-third the depth of the body; the third spine scarcely shorter than the second, but much slenderer; the soft rays gradually decreasing from the first, which is a little shorter than the third spine, to the last but one, the last itself slightly prolonged to form a point backwards. Caudal forked nearly to its base; the lobes much elongated; the upper one, which is a trifle longer than the lower, contained rather more than three times and a half in the entire length. Pectorals narrow and pointed, a little shorter than the head, and contained not quite four times and a half in the entire length; fifth and sixth rays longest. Ventrals attached a little behind the pectorals, and not much more than half their length ; the spine about three-fourths the length of the soft rays, and scarcely stouter than the third spine in the anal : the axillary elongated scale three-fourths the length of the spine. The scales on the body of this species are not materially different from those of the *G. Plumieri* described in the " Histoire des Poissons."

B. 6; D. 9/10 ; A. 3/7 ; C. 17, &c. ; P. 16 ; V. 1/5.

Length 7 inches.

COLOUR.—" White, silvery."—D. The fins are yellowish; the membranes here and there dotted with black : the lobes of the caudal are bordered internally with dusky. I see no trace of the interrupted longitudinal bands spoken of by Cuvier and Valenciennes, neither is there any allusion to them in Mr. Darwin's notes taken from the recent fish.

Habitat, Keeling Island, Indian Ocean.

I do not feel confident as to this species being, any more than the last, identical with that to which I have referred it. It requires an inspection of a large number of specimens in order to ascertain the true value of characters. The present one agrees with what is stated of the *G. Oyena* by Cuvier and Valenciennes, excepting that the second anal spine, which they represent as shorter than the second dorsal spine by one-half, is here shorter by one-third only : also, as mentioned above, there is no appearance of any longitudinal bands. There is no other species in the " Histoire des Poissons," to which it approaches more closely. But comparing it with Rüppell's figure, if this last be scrupulously exact, there are a few other differences besides those already alluded to. Thus the first anal spine in Mr. Darwin's specimen appears shorter in relation to the second, and this last stouter as well as longer. Also the soft rays of this fin gradually decrease, giving a sloping direction to the margin, whereas in Rüppell's figure, all the rays are nearly of the same length, and made equal to the second and third spines. The caudal lobes, likewise, appear longer in Mr. Darwin's specimen. It must be left for others to determine whether these discrepancies are indicative of a specific difference or not. As regards the geographic range of the *G. Oyena*, I know not that there is any thing in this respect to render its identity with the species here described improbable. It inhabits the Red Sea ; and is also said to be common at the Mauritius ;—whence it may very possibly extend as far eastward as the Keeling Islands, where Mr. Darwin's specimen was obtained.

The *Sparus erythrurus* of Bloch (pl. 261) is so extremely unlike the present species both in form and colours, that, except on the authority of MM. Cuvier and Valenciennes, who state that they had seen Bloch's original specimen, no one could have suspected that the figure had been intended for it.

FAMILY—CHÆTODONTIDÆ.

CHÆTODON SETIFER. *Bloch.*

Chætodon setifer, *Bloch,* Ichth. pl. 425. fig. 1.
——————— *Cuv. et Val.* Hist. des Poiss. tom. vii. p. 58.

FORM.—This species is one of those characterized by a prolongation of a portion of the soft dorsal fin. In the present specimen it is the sixth soft ray which is thus prolonged. The total length of this ray, measured from the root, is half the entire length of the fish; and that portion of it which exceeds the adjoining rays is rather more than half. Although the preopercle can hardly be called denticulated, yet there are some faint traces of rudimentary denticulations at the lower angle. The general form, in all other respects, agrees with the descriptions of Cuvier and other authors.

D. 13/24; A. 3/21; C. 17, and 6 short; P. 16, the first short; V. 1/5.

Length 6 inc. 3 lines.

COLOUR.—" Body pale, with narrow dark straight lines which form network: across the eye a black band: posterior half of the body bright orange: upper part of the prolongation of the dorsal fin edged with black, and a round patch of the same."—D. The black ocellus extends from the fifth to the thirteenth ray of the soft dorsal. There is no trace of the four red or yellow streaks said by Cuvier and Valenciennes to cross the forehead from eye to eye; but probably they are effaced by the action of the spirit.
A *second specimen* only differs from the above in being smaller, measuring in length not quite five inches; in having the fifth (instead of sixth) ray in the soft dorsal prolonged; and in the ocellus extending from the fifth to the tenth ray only. In the last two respects it agrees better with the description in the 'Histoire des Poissons.' The filamentous ray terminates in an extremely fine hair, which leads me to think that the extreme portion of this ray in the first specimen has been broken off.

Habitat, Keeling Island, Indian Ocean.

Mr. Darwin's collection contains two individuals of this species procured on coral reefs at the Keeling Islands. As according to his notes made from the recent fish, the posterior half of the body is bright orange, Bloch's figure may not be so much overcoloured as is supposed by Cuvier and Valenciennes, who state that he has represented of a bright red, what ought to be silvery grey and yellow ochre. Perhaps the colours may depend in some measure on the season. Mr. Darwin's specimens were obtained in the month of April.

Genus—STEGASTES.* *Jen.*

Corpus oblongo-ovale, compressum. Caput obtusum. Os parvum, haud protractile. Dentes maxillares omnes incisores, parvi, æquales, contigui, uniseriati; palatini velutini, minuti. Ossa suborbitalia denticulata. Preoperculum margine adscendenti levissimè denticulato. Operculum inerme. Membrana branchialis quatuor-radiata. Pinnæ verticales squamis confertis ferè omnino obtectæ: dorsalis unica, subæqualis, membraná ad apices spinarum parum laciniatá: ventrales radio primo molli elongato. Linea lateralis sub terminationem dorsalis interrupta. Squamæ rostri et verticis parvæ; operculi et corporis magnæ, obliquè dispositæ; omnes levissimè ciliatæ.

This apparently new form will enter into none of the genera established by Cuvier and Valenciennes. The palatine teeth serve to detach it from the *Sciænidæ*, while this character, taken in connexion with the compressed body, and the extreme scaliness of the vertical fins, require that it should be arranged with the *Chætodontidæ*, or at least have a place in that large group to which Cuvier has given the name of *Squammipennes*. It belongs to the second tribe in that family characterized by cutting teeth; and it would seem most nearly allied to *Pime-lepterus*, but it does not approach that genus very closely, and may at once be distinguished from it, by the teeth being without spurs behind, and the dorsal and anal fins being more scaly. From *Dipterodon*, the only other genus in that tribe, it may be known by its undivided dorsal, independently of other marked differences.

But though this genus requires to be arranged with the *Chætodontidæ* on the grounds above mentioned, in all its other characters it comes much nearer that portion of the *Sciænidæ* which have the lateral line terminating beneath the end of the dorsal fin; especially *Pomacentrus*, which it resembles in the general form

* Στεγαστης, tector.

of the head and body, denticulated suborbital and preopercle, unarmed opercle, four-rayed branchiostegous membrane, and in the size and mode of arrangement of the scales on the body. I am not aware that any species of *Pomacentrus* have the dorsal and anal fins so completely covered with scales : but, according to Cuvier and Valenciennes, there is a species of *Glyphisodon*,* to which genus *Pomacentrus* is closely allied, which has these fins almost as entirely scaled, as in the true *Squamipinnati ;* and if so, there is nothing but the palatine teeth which of necessity demands the separation of this new genus from the *Sciænidæ*. These teeth can be distinctly felt upon the vomer, but I am not quite sure from the small size of the fish, and its mouth also being small, whether they exist on the palatines as well. It may be added that this genus shews further itself an affinity to *Glyphisodon*, in the filamentous prolongation of the first soft ray in the ventrals. This character is not, I believe, found in *Pomacentrus*.

In which ever family it is placed, it forms a beautiful connecting link between the two. It is from the Cape Verde Islands.

STEGASTES IMBRICATUS. *Jen.*

PLATE IX. fig. 2.

FORM.—Oblong-oval ; the body much compressed. Greatest depth rather more than one-third of the entire length : head one-fourth of the same. Snout short and obtuse ; the profile rising very obliquely, and forming with the dorsal line one continuous curve. The back is sharp, and appears more elevated than it really is, in consequence of the dorsal fin being thickly coated with scales, and scarcely distinguishable from the body. Ventral line less convex than the dorsal ; the edge of the abdomen somewhat carinated between the ventral and anal fins, but in advance of the former rounded. The upper and under profile meet at the mouth at a right angle. Mouth small, and scarcely at all protractile. Jaws equal ; each with a single row of cutting teeth, which are small, though rather larger below than above, even and closely set, forming a compact series : no secondary teeth behind : vomer rough with minute velutine teeth. When the mouth is closed, no portion of the maxillary is concealed by the suborbital. Eyes round, moderately large, their diameter rather less than one-third the length of the head, placed high in the cheeks, and nearer the end of the snout than the posterior angle of the opercle, the distance from the former being rather less than one diameter. The nostrils consist of a single minute round aperture, about half-way between the eye and the anterior margin of the suborbital. The suborbital has its margin entire as far as the end of the maxillary, at which point it curves backwards and upwards to form a narrow band beneath the eye, and the lower margin of this band is denticulated. The preopercle is likewise denticulated ; but the denticulations, which are principally confined to the ascending margin, are not very obvious, and more

* *G. chrysurus*, Cuv. et Val.

readily felt than seen : the angle at bottom is rounded, and rather exceeds a right angle; a vertical from the angle would form a tangent to the posterior edge of the orbit : the ascending margin is not quite straight, bending slightly inwards a little below the middle. The opercle terminates posteriorly in a very obtuse angle, and shows some indication of two very minute flattened points, which, however, do not project beyond the membrane : from the lowermost of these points the margin of the subopercle passes obliquely forwards to form a continuous curve with that of the interopercle, which is tolerably well developed. Gill-opening of moderate size : the branchial membrane, which apparently has only four rays, has a shallow notch in front, and passes continuously from one side to the other, without being attached to the isthmus.

The lateral line commences at the upper angle of the opercle, and, inclining upwards, runs parallel, not to the dorsal line which can hardly be distinguished, but to the upper edge of the dorsal fin, its distance from which is contained about three times and a half in the entire depth; it terminates a little before the termination of that fin. Cranium, snout, cheeks, pieces of the opercle, the body, and all the vertical fins, covered with finely ciliated scales ; those on the crown and snout small, but those on the opercle and body large; the latter arranged in oblique rows ; about twenty-seven in a longitudinal line from the gill to the caudal, and about fourteen in a vertical one from the dorsal to the ventral line : a scale taken from the row beneath the lateral line, and about the middle of the body, is of an oblong form, its breadth exceeding its length, with the free edge dotted and finely ciliated, the basal margin rather deeply crenated, the crenations separated by seven striæ, which are carried on for only a short way, and do not converge to a fan. The scales on the dorsal and anal fins are small and closely compacted ; those on the former arranged obliquely, but the line of obliquity is in the opposite direction to what it is on the body.

The dorsal fin commences in a line with the posterior angle of the opercle, and occupies a space equalling half the entire length : the height of the spinous portion is nearly uniform, but slightly increases backwards ; between the tips of the spines, the membrane is a little jagged : the soft portion is scarcely more than one-third the spinous in length, but is somewhat higher, terminating upwards in an acute angle ; the longest of the soft rays is about half the depth of the body, the dorsal fin itself not included. The anal answers to the soft portion of the dorsal, which it exactly resembles; it has two spines in front, the first of which is very short, and scarcely more than one-third the length of the second, which itself is shorter than the soft rays ; the second spine is stouter than any of the dorsal spines. These two fins terminate in the same vertical line. The caudal appears to have been square, but the rays are worn at the tips, so that its exact form cannot be ascertained ; it is coated with scales for four-fifths of its length from the base. Between the dorsal and the caudal fins is a space equalling not quite one-third the depth of the body. Pectorals attached a little behind the opercle, and a little below the middle; slightly pointed ; about the length of the head or rather shorter; the first ray only half the length of the second ; fourth and fifth longest ; all the rays, with the exception of the first two and the last two or three, branched. Ventrals attached a little further back than the pectorals ; the first soft ray prolonged into a filament reaching to the commencement of the anal ; the spine is about half the length of the filamentous ray, and about two-thirds that of the second soft ray. Between these fins is an oval lanceolate scale about one-third their length ; and in their axillæ another elongated one, narrower and more pointed than the former, and rather exceeding it in length.

FISII. 65

B. 4 ; D. 12/16 ; A. 2/12 ; C. 15, and 4 short ; P. 21 ; V. 1/5.

Length 3 inches.

Colour.—Not noticed in the recent state. *In spirits*, the whole fish, fins included, appears of a uniform dark brown.

Habitat, Porto Praya, Cape Verde Islands.

The only specimen of this new genus which exists in the collection was taken by Mr. Darwin off Quail Island, in the bay of Porto Praya. It is small, but probably full-sized, or nearly so ; since the greater part of the species of *Pomacentrus*, to which genus it is so strongly allied, average about the same dimensions. Possibly some of the generic characters, which I have given above, may prove hereafter to be merely specific ; but till other species shall have been discovered, their exact value cannot be ascertained.

Family.—SCOMBRIDÆ.

Genus—PAROPSIS. *Jen.*

Corpus altum, rhomboideum, valdè compressum, squamis minutissimis obtectum. Linea lateralis anticè sursum pauló arcuata, per totum longitudinem inermis. Cauda lateribus haud carinatis. Dentes in utráque maxillá uniseriati, tenuissimi, acuti ; in linguá, vomere, et palatinis, velutini brevissimi. Apertura branchialis amplissima, membraná decem-radiatá. Spinæ quinque liberæ loco pinnæ dorsalis primæ ; spiná minutá præeunte reclinatá antrorsum flexá. Dorsalis secunda, æquè ac analis, continua, sine pinnulis falsis: ante analem spinæ duæ liberæ. Pectorales parvæ. Ventrales nullæ. Caudalis profundè furcata, lobis acuminatis, subelongatis.

This new genus belongs to that section of the *Scombridæ* characterized by having a number of short free spines, instead of a first dorsal fin. It is most nearly allied to *Lichia*, especially to the *L. glaucus*, which it resembles in general form, as well as in many of its particular characters. It has the same reclined spine in front of those which represent the first dorsal, and the same two free spines in front of the anal ; also the same form of opercle ; the same deeply-forked caudal, and small pectorals. But it may be at once distinguished from that genus by the absence of ventrals, of which there is not the least trace : the body is also deeper, rhomboidal rather than oval, and more compressed. In all these respects it agrees better with *Stromateus*, which would seem particularly to meet it in those species, such as the *S. candidus* and *S. securifer*, which are represented by Cuvier and Valenciennes as having a number of minute truncated

K

spines before the dorsal and anal fins, and which, by virtue of this character, though in the case of the former the spines are not apparent externally, serve manifestly to re-conduct to the section to which *Lichia* belongs. The discovery of the present genus, therefore, furnishes a more completely connecting link between these two groups.

Rhynchobdella and *Mastacemblus* agree with *Paropsis*, both in wanting ventrals and in having the first dorsal represented by free spines ; but the form of these two genera is so totally different in all other respects, that it is impossible they can be confounded with it.

This new genus is from the east coast of South America.

PAROPSIS SIGNATA. *Jen.*

PLATE XIII.

P. argentea, nitens, summo dorso cœrulescente ; operculo ad angulum superiorem maculâ nigrâ signato ; pinnis pectoralibus maculis duabus in axillis et ad radices radiorum, minoribus.

B. 10 ; D. 5—1/33 ; A. 2—1/35 ; C. 17, et circa ⅘ accessar. ; P. 21 ; V. 0.

LONG. unc. 9.

FORM.— Body very much compressed, of a rhomboidal form, the dorsal and ventral lines rising to an angle at the commencement of the dorsal and anal fins respectively. Head a laterally compressed cone : tail becoming suddenly attenuated before the setting on of the caudal fin, without any keel at the sides. Back sharp and elevated ; the greatest depth contained not more than two and a half times in the entire length : thickness only one-fifth of the depth. The length and depth of the head are equal, each being half the depth of the body. The upper and under profile meet at the extremity of the snout at nearly a right angle, the former falling in a very regular curve from the commencement of the dorsal fin. Mouth moderately large, the commissure reaching to beneath the eye, with the lower jaw projecting and of considerable strength and thickness. In each jaw a single row of very fine sharp teeth. The tongue, which is of a triangular form, free at the tip, and pointed, is rough, with some extremely fine closely shorn velutine teeth : a small triangular patch of these last teeth on the front of the vomer, and a narrow row on each palatine : pharyngeans with rather stronger teeth. The intermaxillary is very slightly protractile. The maxillary reaches, when the mouth is closed, to a vertical from the posterior part of the orbit : it is very visible from without, having only its anterior portion concealed by the suborbital, and being much dilated at its posterior extremity, which is in shape somewhat securiform. Eyes above the middle of the cheek, and nearer the end of the snout than the posterior margin of the opercle ; their diameter rather more than one-fifth the length of the head : the suborbital forms a narrow band beneath each. Nostrils half-way between the eyes and the end of the snout ; the anterior orifice round ; the posterior, which is the larger one, oval. Preopercle with the ascending margin nearly vertical ; the angle at bottom rounded. The opercle and subopercle together present a rounded margin posteriorly,

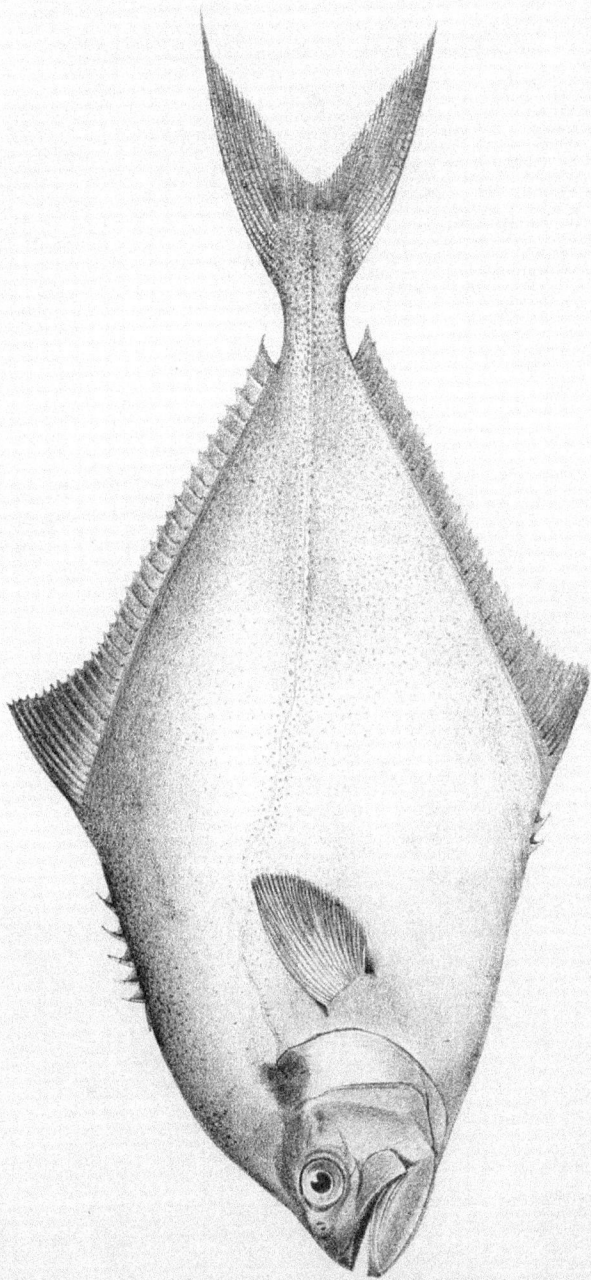

Parapsis signata Nat. size

N. Hawkins del.t

.

though at the upper portion there are two small blunt points distinguishable by the finger, between which there is a very shallow notch: the line of separation between these two bones descends obliquely forwards to a little above the rounded angle of the preopercle, where it meets the line of the interopercle, which descends obliquely backwards:* all the margins of the opercular pieces entire. Gill opening very large, the aperture reaching to beneath the anterior margin of the eye: the membranes, each of which has as many as ten rays,† cross a little over each other, and are not united to the isthmus.

Snout, jaws, and cheeks, as well as the several pieces of the gill-cover, without scales: ‡ body covered with extremely minute ones, of an oval form, longer than broad, marked with concentric circles, and entire on the margin. The lateral line is slightly arched above the pectoral, and its course a little undulating, but it descends gradually to near the middle of the body, whence it runs straight to the caudal.

The first dorsal is represented by five short free spines, each capable of separate motion, and each furnished with its own membrane; in advance of them is a somewhat smaller reclined spine with its point directed forwards: the first erect spine is above the middle of the pectoral, and distant from the end of the snout nearly one-third of the entire length. Beyond the five free spines, and immediately before the commencement of the second dorsal is another small spine closely pressed down, and almost concealed beneath the skin, pointing backwards. The second dorsal, which has also at its anterior edge a small spine one-third the length of the first soft ray, commences at the middle point of the entire length, caudal excluded. The general form of this fin is similar to that of the genus *Lichia*, long, with the anterior portion elevated; the greatest height about one-fourth of the depth, or hardly so much. The anal answers exactly to the second dorsal in form and extent, and is preceded by two free spines, separated from it by a small space, besides a longer one at its anterior edge. Caudal forked nearly to the base, where there are a number of minute scales; the lobes equal, pointed, and moderately elongated, each contained about four times and one-third in the entire length. Pectorals attached at about the middle of the depth, a little behind the opercle; of a somewhat triangular form, small, their length not much exceeding half that of the head. No trace of ventrals whatever.

COLOUR.—" Uniform bright silvery, the ridge of the back bluish: a black patch on the gill-cover, and another under the pectoral fin."—D.—The first of the patches alluded to by Mr. Darwin is very conspicuous, and is situate at the upper angle of the opercle, immediately in advance of the commencement of the lateral line. The second may be described as consisting of two distinct spots; one at the root of the upper rays, and completely in the axilla; the other, a small one of an elongated form, immediately beneath the lowest ray, and partly visible without raising the fin. The elevated portion of the second dorsal is also dusky, and a faint edging of this colour runs for a short way along the margin of this fin. The anal is pale.

Habitat, Northern coast of Patagonia.

* This part is exactly as described by Cuvier and Valenciennes in the *Lichia Amia*, to which genus the present one is so nearly allied.

† *Lichia Amia* is represented as having nine; and this forms another mark of affinity between the two genera.

‡ There are scales on the cheeks in *Lichia*, according to Cuvier and Valenciennes, but I see no appearance of them in this genus.

I have termed this species *signata*, in reference to the black patch on the opercle, which is a conspicuous character. The only specimen in the collection was obtained by Mr. Darwin at Bahia Blanca, on the coast of North Patagonia.

1. CARANX DECLIVIS. *Jen.*

PLATE XIV.

C. corpore elongato, altitudine quintam, capite quartam partem longitudinis æquante; maxillâ inferiore longiore; lineâ laterali infra quintum radium dorsalis secundæ subito declivi, per totam longitudinem armatâ, laminis 82 altioribus quum longis, ubique æqualibus; spinâ reclinatâ ante pinnam dorsalem parvâ, mucrone tamen nudato; pectoralibus ultra pinnulam analem, et prope ad analem ipsam, pertingentibus.

B. 7 ; D. 8—1/35 ; A. 2—1/30; C. 17, &c. ; P. 21 ; V. 1/5.

LONG. unc. 7. lin. 10.

FORM.—Rather more elongated than the *C. trachurus* of the British seas. Greatest depth one-fifth of the entire length: head one-fourth of the same : thickness about half the depth. Diameter of the eyes a little less than one-third the length of the head. Lateral line bending downwards more suddenly, and at a more backward point than in that species. The bend commences in a line with the fifth ray of the second dorsal, and is entirely comprised within a space equal to that occupied by four fin rays,* so that opposite the ninth ray it again advances in a horizontal line. The posterior portion about equals in length the anterior, the bend being included in this last. The laminæ which protect the lateral line, and which extend throughout its whole length, are well developed, and everywhere of the same breadth ; this breadth equalling nearly, but not quite, one-third the depth of the body. In number they are eighty-one or eighty-two ; of which the last thirty-eight or forty, forming the posterior portion of the line, have keels terminating backwards in sharp spines : these spines are at first small and inconspicuous, but gradually increase in size as they advance towards the thinnest part of the tail, where they are sharpest and most developed.

In most of its other characters this species so closely resembles the *C. trachurus*, as to render a detailed description unnecessary. The reclined spine before the first dorsal, however, is smaller, though the point is sharp and exposed : also the number of rays in the second dorsal and anal is greater by five in each fin. The length of the second dorsal is two-and-a-half times that of the first. The pectorals are long, narrow, and pointed ; a little shorter than the head, or rather less than one-fourth of the entire length; when laid back, they reach beyond the anal finlet, and very nearly to the commencement of the true anal itself.

COLOUR.—Not noticed in the recent state. So far as can be judged from a specimen in spirits, the colours appear to have been similar to those of the *C. trachurus;* and there is the same black spot on the upper part of the opercle.

* In the *C. trachurus*, the bend begins in a line with the commencement of the second dorsal, and from its more gradual obliquity, extends over a space equal to that occupied by nine fin-rays.

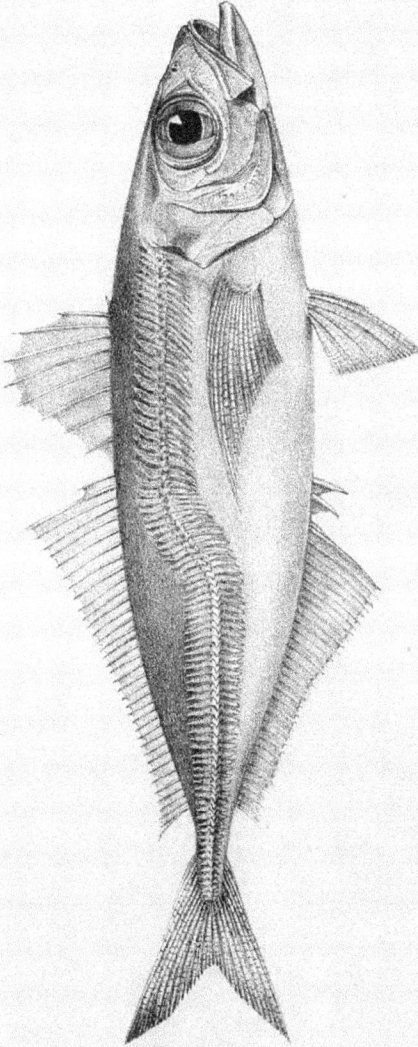

Caranx declivis Nat Size

W. Harrison del.

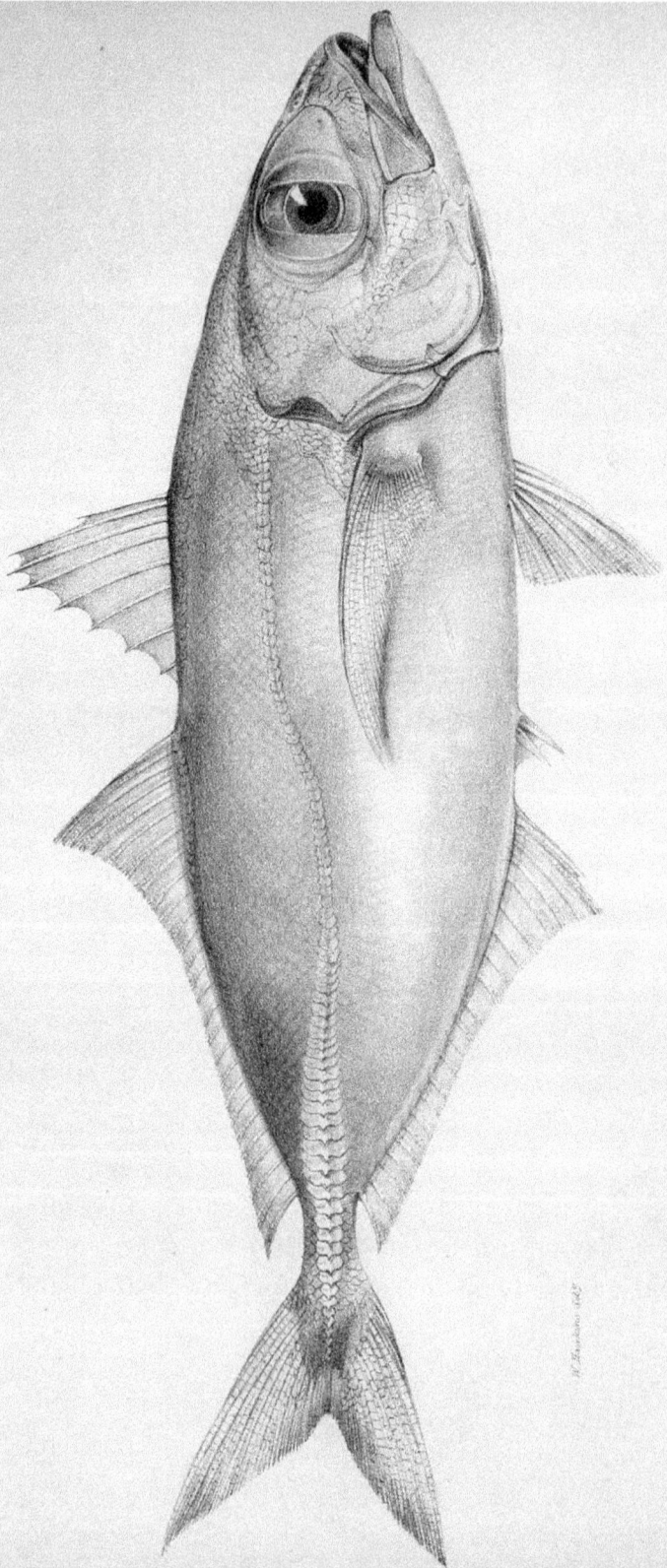

Caranx lutus. Nat. size.

H. Richardson del.t

Habitat, King George's Sound, New Holland.

Cuvier and Valenciennes have noticed several variations of form occurring in different specimens of the *C. trachurus*, from different seas, which they have not ventured to raise to the rank of species. That the one here described is entitled, however, to this distinction, I can hardly entertain a doubt. The suddenness of the bend in the lateral line, and the more backward point at which the bend commences; the larger number of laminæ which protect it; and also the larger number of rays in the second dorsal and anal; all seem to indicate a specific difference. Whether it be identical with any of the varieties noticed by them is uncertain; but it seems to be distinct from the only one they speak of as having been received from New Holland, in which the number of laminæ did not exceed seventy-three. I have called it *declivis*, in reference to the character of the lateral line above alluded to. It was obtained by Mr. Darwin in Princess Royal Harbour, in King George's Sound.

2. CARANX TORVUS. *Jen.*

PLATE XV.

C. corpore crassiusculo, subelongato; altitudine vix quartam partem longitudinis æquante, capite quartam superante; maxillâ inferiore longiori; oculis magnis; suborbitalibus venis nonnullis subparallelis obscuris notatis; lineâ laterali parum deflexâ, anticè squamis parvis inermibus, posticè laminis carinatis 35 vel 36 tectâ; spinâ reclinatâ ante pinnam dorsalem sub cute occultâ; pectoralibus longis, falcatis, ad initium pinnæ analis prope pertingentibus.

D. 8—1/26; A. 2—1/22; C. 17, &c.; P. 21; V. 1/5.

LONG. unc. 11. lin. 9.

FORM.—Thicker and deeper in the body than the *C. trachurus*. The greatest depth a little less than one-fourth of the entire length; the thickness exceeding (but by a very little) half the depth. Head large; its length a little more than one-fourth of the entire length; its height or depth, taken in a line forming a tangent to the posterior part of the orbit, less than its own length by two-sevenths. Eyes large; their diameter very nearly one-third the length of the head; partially covered at the sides by two fatty membranous veils, as in several other species. The ventral line of the body is rather more curved than the dorsal, and the upper profile in like manner a little more approaching to rectilineal than the lower. The lower jaw a little the longer, and ascending to meet the upper. Maxillary reaching not quite to beneath the middle of the eye; its extremity truncated in the form of an arc, with the curvature inwards. In each jaw a single row of very fine, minute, closely set teeth; two small patches on the anterior extremity of the vomer, a band on each palatine, and one on the tongue, all closely shorn velutine. Suborbital, on each side of the extremity of the snout, marked with several nearly parallel dark-coloured veins. Preopercle with the angle very much rounded; the limb broad, slightly striated or veined, and not separated from the cheek by any salient ridge. The other pieces of the gill-cover taken together are bounded posteriorly by a sinuous and very irregular margin, the notch in the bone at the upper part of the opercle being nearly semicircular, beneath which is an obtuse point, whence the obliquely descending margin first slopes slightly inwards, then passes outwards to form another blunt point lower down, then slopes inwards again. The course taken by the margin of the membrane in some measure follows that of the bone, but the sinuosities and salient angles are more rounded. Cheeks and opercle scaly, as well as the cranium and forehead between the eyes.

The lateral line does not deviate very much from rectilineal. The deflection, such as it is, may be said to commence in a line with the termination of the first dorsal, and to end beneath the first third of the second dorsal. Up to this point, the scales which cover it are small and round; but they then begin gradually to enlarge, and to assume a keel terminating posteriorly in a short spine : these scaly laminæ continue increasing in size till they arrive beneath the last quarter of the fin, where they are most developed ; none of them, however, are very large, and even here they do not extend over the whole breadth of this part of the tail, nor their own breadth exceed one-eighth of the greatest depth of the body. After passing the dorsal and anal fins, they rapidly diminish as they approach the caudal. The entire number of laminæ may be set at thirty-five or thirty-six; but as it is difficult to fix the exact point where they commence, it will vary according as the computation is made more or less in advance. The anterior portion of the lateral line, bend included, is a little longer than the posterior.

The reclined spine in this species is entirely concealed beneath the skin. The pectorals are long and falcate, terminating in a sharp point : their length nearly equals that of the head, or about one-fourth of the entire length : when laid back, they reach over the anal finlet, and very nearly to the commencement of the true anal. The ventrals are attached a little behind the pectorals, and are only half as long. The other fins are much as in the other species of this genus. The height of the anterior part of the first dorsal equals exactly half the depth. The lobes of the caudal are one-fifth of the entire length.

Colour.—Not noticed in the recent state. In spirits ; silvery on the abdomen and lower half of the sides, passing above the middle, and on the back, into pale lead blue, tinged with gray and brownish : fins pale greyish brown. No conspicuous markings, except the usual spot on the notch of the opercle, which, however, is small, and confined entirely to the membrane.

Habitat, Tahiti.

This species belongs to the second section adopted by Cuvier and Valenciennes in this genus ; or that in which the form of the body resembles that of the *C. trachurus*, but in which the laminæ on the lateral line only extend over the posterior portion, the anterior being smooth and simply covered with small scales. But it will not exactly accord with any of the species described by those authors. It seems to approach most nearly the *C. Plumieri;* but though the eyes are of considerable size, they are not quite so large as they are represented to be in that. There seem, in fact, to be several species characterized by large eyes. Spix and Agassiz have figured one from America under the name of *C. macrophthalmus;* and under the same name Ruppell has figured another from the Red Sea. Both these, however, appear likewise different from the one here described, at the same time that their different geographic range renders their identity *à priori* improbable. The present one was taken by Mr. Darwin at Tahiti.

3. CARANX GEORGIANUS. *Cuv. et Val.*

Caranx Georgianus, *Cuv. et Val.* Hist. des Poiss. tom. ix. p. 64.

FORM.—Of an oval compressed form, with the back elevated. Greatest depth one-third of the entire length, caudal excluded : thickness not half the depth : head one-fourth of the entire length, caudal included. Profile ascending obliquely, and in nearly a straight line, to meet the dorsal curve. Upper jaw a little the longer. The maxillary, which is truncated and cut nearly square at its posterior extremity, not quite reaches to beneath the anterior margin of the orbit. In each jaw a row of about thirty-five teeth, which are small, somewhat cylindrical, set regularly, nearly equal, and rather blunt at the point; very little trace of any secondary row, simply four or six smaller ones behind those in the middle of the upper jaw, and perhaps in the lower also, but they are not very obvious. A triangular patch of velutine teeth on the vomer, and a narrow band of the same on each palatine; also on the tongue : these last, however, very closely shorn. Eyes a little above the middle of the cheek, but exactly half-way between the end of the snout and the posterior margin of the opercle ; their diameter one-fourth the length of the head. Preopercle rounded at the angle; its limb separated from the cheek by a slight but not very salient ridge. Opercle with the notch at the upper part not very deep; the obliquely descending margin straight.

The lateral line follows the curvature of the back until it arrives beneath the middle of the second dorsal, at which point it becomes straight, and the scales gradually pass into carinated spinous laminæ. These laminæ, however, are very little developed anteriorly to the last quarter of that fin ; and even beneath the end of it, where they are largest, they do not extend over more than half the breadth of the tail, nor does their own breadth exceed one-seventeenth of the greatest depth of the body. The number of them is from twenty to twenty-five, according to the point at which the reckoning commences, the transition from the scales to the laminæ being very gradual. The pectorals are falcate and sharp-pointed, and one-fourth of the entire length, caudal included. The height of the anterior part of the dorsal is contained two and a half times in the depth. The lobes of the caudal are contained four times and three-quarters in the entire length.

D. 8—1/27 ; A. 2—1/24 ; C. 17, &c. ; P. 20 ; V. 1/5.

Length 7 inches 6 lines.

COLOUR.—Not noticed in the recent state. The colour of the back and upper part of the sides appears to have been bluish grey, with steel and other reflections, and was probably very brilliant in the living fish : belly silvery. No markings, except a conspicuous black spot on the upper part of the opercle.

A second specimen.—Differs in no respect from the above, excepting in having one ray less in the second dorsal and anal fins.

Habitat, King George's Sound, New Holland.

I entertain not the least doubt of this species being the *C. Georgianus* of

Cuvier and Valenciennes ; but as the notice of it in the " Histoire des Poissons" is extremely brief, I have deemed it advisable to annex a detailed description. Both Mr. Darwin's specimens are from King George's Sound, where the species was first discovered by MM. Quoy and Gaimard.

Seriola bipinnulata. *Quoy et Gaim.*

Seriola bipinnulata, *Quoy et Gaim.* Voyage de l'Uranie (Zool.) p. 363, pl. 61. f. 3.
———————————— *Cuv.* Regne An. (2d Edit.) tom. ii. p. 206.

Form.—Elongated, and fusiform. Greatest depth contained four times and a half in the length, measuring this last to the base of the caudal fork. Head four times and a quarter in the length : depth of the head not quite once and three-quarters in its own length ; the cheeks nearly vertical. Snout pointed : profile straight, and but slightly falling. Lower jaw a little longer than the upper, the commissure reaching to beneath the orifices of the nostrils : maxillary very conspicuous, and greatly dilated at its posterior extremity. A band of minute velutine teeth in each jaw, broadest in front ; a disk of similar teeth on the vomer, and a band on each palatine. Eyes large ; their diameter one-fifth the length of the head ; situated a little above the middle of the cheek, and a little nearer the end of the snout than the posterior margin of the opercle ; exactly two diameters between the eye and the end of the lower jaw. The nostrils consist of two small, round, closely approximating orifices, the anterior one partially covered by a membrane ; situated rather nearer the eye than the extremity of the snout. Preopercle with the ascending margin vertical, and the angle at bottom rounded ; the limb very broad, and marked with veins, and between the veins, along the basal margin, with fine striæ. The rest of the pieces of the gill-cover, taken together, present a rounded and regularly curved outline posteriorly ; the line of separation between the opercle and subopercle ascends obliquely backwards from a point about two-thirds down the posterior margin of the preopercle ; that between the subopercle and the interopercle (which last is well developed) passes downwards and backwards, forming an angle of about 45° with the axis of the body. Branchial aperture large ; the membrane deeply cleft. Snout, jaws, and pieces of the opercle, smooth and naked ; cheeks scaly, the scales on the upper part of the cheek, between the eye and the upper angle of the preopercle, being of a narrow pointed form. The scales on the body are of a moderate size, oval, marked with fine concentric circular striæ, with a fan of coarser diverging striæ on their concealed portion. The lateral line is smooth throughout its length, and runs nearly straight from the upper angle of the opercle to the caudal, its course being a little above the middle.

The first dorsal commences at about one-third of the entire length, measuring this last as before : it is low and inconspicuous, consisting of only six weak spines, of which the third and fourth are somewhat the longest, but whose length is less than one-fifth of the depth of the body. The length of the fin itself is rather less than half the depth. Second dorsal closely following, and much longer ; of the form usual in this family, with the anterior portion elevated and somewhat triangular, but beyond the ninth ray low and even : its spine half the length of the first soft ray : its greatest elevation contained about two and a-half times in the depth. The last two rays of this fin are broke away from the rest, with an intervening space, to form a spurious fiulet, and are rather longer, the last especially, than those which precede. The anal commences opposite the fourteenth ray of the second dorsal, and is similar in form to that fin, but of course shorter, and also less elevated at its anterior extremity : finlet and the intervening space exactly corresponding. Caudal deeply forked ; the lobes very long and pointed, each equalling nearly one-fourth of the entire length ; the middle rays not one-fourth the length of the lateral ones. Pectorals attached a little below the middle ; in length a little exceeding half that of the head. Ventrals about the same size as the pectorals, but attached a little further back. A slight elevation at the sides of the tail, but no distinct keel, properly so called.

D. 6—1/24—I ; A. 1/16—I ; C. 17, &c. ; P. 20 ; V. 1/5.

Length 18 inc. 3 lines.

COLOUR.—" Band on the side azure blue ; above a duller greenish blue ; beneath two greenish metallic stripes : lower half of the body snow white."—D. No trace of the longitudinal stripes remains in the dried skin.

Habitat, Keeling Island, Indian Ocean.

A tolerably exact figure of this species occurs in the Zoological Atlas of Freycinet's Voyage, but I can find no notice of it in the "Histoire des Poissons" of Cuvier and Valenciennes. Although referred by Cuvier in his "Regne Animal" to *Seriola*, it rather departs from that genus in some of its characters. Independently of the spurious finlets in the dorsal and anal fins, which separate it from all the other species, I see no trace of any reclined spine before the first dorsal, nor of two free spines before the anal ; in both which respects *Seriola* is said to resemble *Lichia*. Possibly, however, as Mr. Darwin's specimen is a dried skin, these characters may have been destroyed in the process of preparation. And to the same cause, perhaps, is to be attributed the circumstance of my not being able to observe more than one spine in the true anal, Quoy and Gaimard mentioning two. On the other hand, these naturalists appear to have overlooked the narrow pointed scales on the upper part of the cheeks, which are of a different character from the scales on the body.

Mr. Darwin's specimen of this species was obtained at the Keeling Islands. The one figured in Freycinet's Voyage was procured at Papua or New Guinea. It probably, therefore, has a considerable range over the Indian Ocean.

PSENES —— ?

Psenes leucurus, *Cuv. et Val.?* Hist. des Poiss. tom. ix. p. 197.

Mr. Darwin's collection contains two individuals of a species of *Psenes*, in reference to which his notes state that they were taken in Lat. 17° 12′ S., Long. 36° 33′ W., a hundred and twenty miles from the nearest land above water, though shoals were considerably nearer. They do not measure more than one inch eight lines in length ; and from their small size, and their not being in a very firm state of preservation, it is hardly possible to say whether they are new or not. In form, they differ but little from the *P. cyanophrys* of Cuvier and Valenciennes : still they are evidently not that species, and one point of difference consists in the lateral line, which terminates beneath the end of the second dorsal, and is not carried on to the caudal, as represented in the figure of the above species in the " Histoire des Poissons :" the eye too appears rather larger ; the forehead is hardly so much elevated, and the pectorals are shorter than the head. Perhaps it may be the *P. leucurus* of the above authors ; though this species is from the Indian seas, so that its range must be considerable if the same. The description of the *P. leucurus* in the " Histoire des Poissons" is too short to determine this point. It is said to have been so named on account of its whitish tail, all the other fins being black. In the present species, the fins are likewise black, or at least dusky, except the caudal, which Mr. Darwin's notes, taken from the recent fish, state to have had " a pink tinge." In the same notes it is added,—" belly silvery white mottled with brownish black ; sides bluish with dusky greenish markings ; iris yellow, with dark blue pupil." The fin-ray formula is as follows :

D. 10—1/27 ; A. 3,27 ; C. 17, &c. ; P. 17 or 18 ; V. 1,5.

Though these specimens are small, they have the appearance of being nearly full-sized. Cuvier and Valenciennes state that their specimens of the *P. leucurus* do not exceed two inches in length.

STROMATEUS MACULATUS. *Cuv. et Val.?*

Stromateus maculatus, *Cuv. et Val.* Hist. des Poiss. tom. ix. p. 296.

FORM.—General form so extremely similar to that of the *S. Fiatola* of the Mediterranean as to preclude the necessity of a detailed description. Greatest depth one-third of the length : head one-fifth of the same. Number of rays in the dorsal and anal fins somewhat greater than in the *S. Fiatola.* The height of the dorsal also a little greater, being contained about three times and a half in the depth: the fifth and sixth soft rays longest. Fleshy part of the tail more slender. Pectorals about the length of the head.

B. 6; D. 7/41 ; A. 5/40 ; C. 17, besides several short ; P. 23 ; V. 0.

Length 8 inches 6 lines.

COLOUR.—"Silvery blue above, with regular circular leaden spots."—D. The spots are small, and of nearly equal size: they prevail from the back downwards to about the middle of the depth, and advance a little on the base of the dorsal fin. The arrangement of them is much as described in the " Histoire des Poissons."

Habitat, Chiloe, West Coast of S. America.

It is just possible that this may not be specifically the same as the *S. maculatus* of Cuvier and Valenciennes, but it comes so extremely near that species that I do not feel authorised in describing it as distinct without seeing more specimens. It is stated by the authors above mentioned, that the fin-ray formula of the *S. maculatus* is the same as that of the *S. Fiatola:* in the specimen here described, the number of rays in the dorsal and anal fins appear to me somewhat greater ; but as the spines of these fins are very minute at their commencement, and not readily counted, nor very distinguishable from the soft rays, perhaps the discrepancy may arise from a difference in the mode of computation. What is more to be noted is, that the spots, although they agree in form and mode of arrangement, are said by Mr. Darwin, in his notes taken from the recent fish, to have been "leaden;" whereas it is stated in the "Histoire des Poissons" that they are "yellow." Perhaps they may vary in colour according to the period of the year. There is likewise a difference in locality as regards latitude. The *S. maculatus* is said to be common in the market at Lima, and to have been brought also, both by M. D'Orbigny and M. Gay, from Valparaiso. Mr. Darwin's specimen, however, was taken as far south on the western coast of S. America as Chiloe.

Mr. Darwin's collection contains another specimen, either of the same species as that described above, or one so extremely similar to it as not to be distinguishable in the case of this specimen, which is in too bad preservation to admit of an accurate description of it being given. The following, however, are Mr. Darwin's notes taken from the recent fish:—

Colour.—" Whole body silvery; upper part of the back iridescent blue, lower greenish ; spotted with coppery-lead circular patches."—D.

This specimen measures ten inches and a half in length. It will be observed that the colour of the spots is still said to have been " lead," though inclining to coppery. It was not taken at the same place as the other, but at Port St. Julian, in central Patagonia ; if therefore they are both referable to the *S. maculatus*, this species will have been proved to have a wide range in point of latitude, and also to occur on both sides of the S. American Continent, which is remarkable, considering that it is found so high up the western side as Lima.

Family.—TEUTHYDIDÆ.

1. Acanthurus triostegus. *Bl. Schn.*

Acanthurus triostegus, *Cuv. et Val.* Hist. des Poiss. tom. x. p. 144.
———— Hirudo, *Benn.* Fish of Ceyl. pl. xi.

This species, which appears to be well known, and to have been described by several authors, was found by Mr. Darwin on coral reefs at the Keeling Islands. Cuvier and Valenciennes observe that it has a wide range through the Indian and

Pacific Oceans. Mr. Darwin's specimen agrees in every respect with the description in the "Histoire des Poissons," except in having one ray more in the anal fin : its length is not quite five inches.

2. ACANTHURUS HUMERALIS. *Cuv. et Val.*

Acanthurus humeralis, *Cuv. et Val.* Hist. des Poiss. tom. x. p. 170.

FORM.—General form oblong-oval. Greatest depth just behind the insertions of the pectorals ; contained exactly twice in the length of the oval of the body (measuring this last from the end of the snout to the base of the caudal spine), and three times in the entire length (measuring this last to the extremities of the lobes of the caudal fin.) Profile convex before the eyes, whence it descends nearly vertically to the mouth. Height of the head a little exceeding its own length. Eyes very high in the cheeks, and in front of each a grooved line passing horizontally forwards towards the nostrils ; which last consist of two small round orifices, the anterior one larger than the other, and partially covered by a membranous flap. There are seventeen teeth in the upper jaw, and sixteen in the lower : those above have the cutting edges crenated, and likewise the lateral edges for nearly half way down ; this most observable in the middle ones, in which the crenations amount to eight or ten in number : those below similar, but with the crenations not quite so numerous, and in some of the teeth at the sides of the jaw almost confined to the cutting edges. Scales minute ; those taken from the middle of the body appear of an oblong form, their apical portions dotted, and ciliated with from twelve to eighteen very minute denticles, their surface marked with extremely fine delicate striæ, not distinguishable without a strong lens.

The lateral line follows the curvature of the back, at about one-fifth of the depth. The spine on the sides of the tail is strong, and sharp-pointed, and very slightly bent. No reclined spine before the dorsal. Both the fin just mentioned and the anal have their soft portions terminating posteriorly in rather an acute angle : also both have a scaly membrane at the base, and rows of minute scales between the soft rays extending for about one-third or more of their length. The first ray of the anal is very minute, and so much concealed in the skin as to be easily overlooked. The filaments of the caudal are sharp-pointed, and extend as far again as the middle rays : the upper one rather longer than the other. The pectorals are contained three times in the length of the oval of the body. Ventrals attached a little further back, sharp-pointed, and terminating in the same vertical line with the pectorals, both being laid back.

D. 9/23 ; A. 3/23 ; C. 16, &c. ; P. 16 ; V. 1/5.

Length, to the end of the caudal lobes, 7 inches.

COLOUR.—The colours appear to have been exactly as described in the " Histoire des Poissons." Mr. Darwin's notes taken from the recent fish state, "splendid verditer blue and green ;" but do not enter into the details of the markings.

Habitat, Tahiti.

Obtained at Tahiti, where it had been previously found by MM. Lesson and Garnot. Mr. Darwin's specimen accords with the characters given by Cuvier and Valenciennes, except in having two soft rays less in the dorsal, and one less in the anal. Their description, however, is not very detailed.

FAMILY.—ATHERINIDÆ.

1. ATHERINA ARGENTINENSIS. *Cuv. et Val.?*

Atherina argentinensis, *Cuv. et Val.* Hist. des Poiss. tom. x. p. 350.

FORM.—Depth exactly one-sixth of the length, measuring this last to the end of the middle caudal rays. The length of the head is five and a-half times in the same, measuring this last to the end of the caudal lobes. Thickness of the body at least half the depth. Head broad and flat, its breadth across the crown behind the eyes equalling three-fourths of its depth. Snout rounded horizontally, but sharp vertically. The profile is perfectly horizontal; and one uniform straight line extends from the extremity of the upper jaw to the commencement of the second dorsal. Ventral line swelling a little outwards, with its greatest curvature about the middle. Upper jaw a very little longer than the lower, which ascends to meet it at an angle of 45°: gape not reaching more than half-way to the eye, at first horizontal, afterwards deflexed. In each jaw two rows of teeth, stronger and more developed than usual in this genus, widely asunder, and at irregular intervals: in the upper jaw these two rows are equal; in the lower the outer row is stronger than the inner: the outer row above contains about thirty-two or thirty-three teeth; that below twenty-six or twenty-eight: no teeth on the tongue, and scarcely any that can be seen on the vomer or palatines, though a slight roughness can be felt on the last two. Eyes moderately large; their diameter a very little less than one-fourth the length of the head; situate a little in advance of the middle point, and also a little above the middle of the depth. Cheeks and gill-covers scaly. Form of the scales of the body, as well as the number of longitudinal rows, exactly as stated by Cuvier and Valenciennes. The same may be said of the lateral line, and the situation of the dorsal fins. The second dorsal and anal terminate in the same vertical line. Pectorals exactly the length of the head. Ventrals attached immediately below the tips of the pectorals. Breadth of the silver band, which runs straight along the middle of the sides, exactly one-fifth of the greatest depth of the body.

D. 5—1/9; A. 1/19; C. 17, &c.; P. 15; V. 1/5.

Length 8 inches.

COLOUR.—"Silvery, with a silver lateral band: above bluish grey."—D. In spirits, it appears greenish brown, becoming deeper above the silver band and on the ridge of the back: the free margins of the scales are finely dotted with black: the rays of the caudal have been worn at the tips, but there is a trace of the dusky edging noticed by Cuvier: the pectorals are also stained with dusky.

Habitat Maldonado.

I conceive there is but little doubt of this being the *A. Argentinensis* of Cuvier and Valenciennes; but as the description in the "Histoire des Poissons" is short, I have thought it advisable to give a more detailed one of the above specimen. Mr. Darwin took it at Maldonado, where he states that it is very common, adding that it is sometimes found in brackish water. M. D'Orbigny had also found it previously at the mouth of the Rio Plata.

2. ATHERINA MICROLEPIDOTA. *Jen.*

PLATE XVI. Fig. I. Nat. size.

Fig. 2. *a, b.* Magnified scales.

A. gracilis; corporis altitudine partem octavam, capite quintam, longitudinis æquante: oculis mediocribus: maxillis sub-æqualibus, parum protractilibus; commissurá primum horizontali, deinde paulo deflexá, haud oculos attingente: dentibus velutinis, serie externá supra subtusque fortiori: dorsali primá omnino pone ventrales reclinatas locatá: squamis parvis, seriebus longitudinalibus octodecim ad minimum dispositis.

D. 5—1/11 ; A. 1/17 ; C. 17, &c.   P. 15; V. 1/5.

LONG. unc. 4.

FORM—More slender and elongated than the generality of the species in this genus. Greatest depth not more than one-eighth of the entire length. Head one-fifth of the same. Greatest thickness (in the region of the gills) equalling two-thirds of the depth, or rather more. Ventral line of the body scarcely more curved than the dorsal. The upper profile falls, though very slightly, from the nape to the mouth, and the lower profile inclines upwards to meet it at about the same degree of curvature. Head broad, its breadth across the crown nearly equalling its depth. Snout horizontally rounded. Jaws nearly equal; not so protractile as in some other species: the commissure of the lips at first horizontal, but posteriorly inclining a little downward, and scarcely reaching more than half-way to the eye. In each jaw two rows of slender very distinct teeth, with traces of a third or even fourth row above, towards the middle: outer row longest and most conspicuous, consisting, in the upper jaw, of from forty-five to fifty teeth ; in the lower of scarcely more than twenty-five. No teeth that can be seen on the vomer or palatines, though a slight roughness can be felt on both. Eyes of moderate size; their diameter rather more than one-fourth the length of the head ; almost entirely before the middle, as well as above it: space between the eyes flat, and exceeding the diameter by about one-third : a slightly elevated line on each side of this space, but no other conspicuous sculpture. Opercle with the descending margin sloping obliquely forwards.

Crown, cheeks, and gill-covers scaly, the scales on the crown extending as far as the eyes. Scales on the body small, the number of longitudinal rows amounting to eighteen or twenty : in form nearly square, the length a little exceeding the breadth, the superficies marked with numerous very distinct concentric lines, the basal half with a fan of from four to six deeper cut striæ, the basal margin rather sinuous, and obsoletely crenate where the striæ meet it. No lateral line very distinguishable.

First dorsal small and delicate, commencing exactly at the middle point of the entire length, measuring this last to the bottom of the caudal fork, and in a line with the tips of the ventrals, these last fins being laid back. Space between the first and second dorsals a little exceeding two-thirds of the depth of the body. Length and height of the second dorsal equal

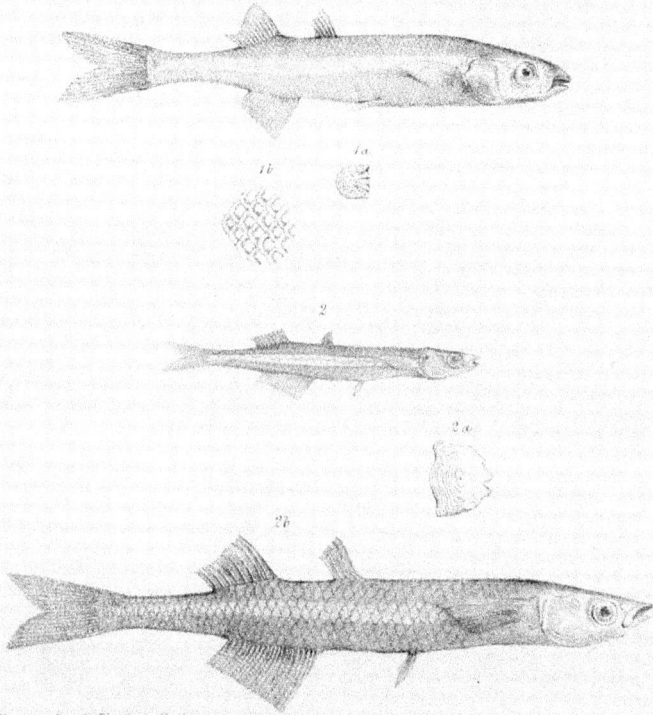

Drawn from Nature on Stone by Waterhouse Hawkins

1 Atherina Microlepidota.   Nat. Size
1a. 1b.   magnified Scales
2 Atherina incisa   Nat. Size
2a.   magnified Scale
2b.   Twice Nat. Size.

to each other, and also to the space just alluded to. From the end of the second dorsal to the commencement of the caudal is exactly one-sixth of the entire length. The posterior half of the anal nearly answers to the second dorsal, but the two fins do not terminate exactly in the same line, the dorsal extending a little the furthest. Caudal forked for about half its length. Pectorals about two-thirds the length of the head. Ventrals attached at a point beyond the extremity of the pectorals. Breadth of the silver band about one-fifth the depth of the body.

COLOUR.—Not noticed in the recent state. In spirits, the back and sides above the silver band are brownish, with the contour of each scale marked out by black dots. All below the band appears to have been silvery. The band itself is not very brilliant. Both the dorsals, as well as the caudal, are dusky: anal and ventrals pale.

A *second specimen* does not differ from the above, excepting slightly in the fin-ray formula, which is as follows:

D. 5—1/10 ; A. 1/15 ; &c.

Habitat, Valparaiso.

This species was found by Mr. Darwin at Valparaiso in fresh water, in the month of August. It would seem to be nearly allied to the *A. laticlavia* of Cuvier and Valenciennes, brought from the same locality by M. D'Orbigny ; but, judging from the short description in the "Histoire des Poissons," it is more elongated, and has the head longer in relation to the depth of the body ; also has the silver band narrower. In the *A. laticlavia*, the head is said to be equal to the depth, and to be contained six times in the entire length ; the breadth of the silver band to be greater than in any other species. In the *A. microlepidota*, the depth is one-eighth and the head one-sixth of the length : the silver band not broader than in the *A. argentinensis* and some others. The colouring also of the fins appears different in the two species.

3. ATHERINA INCISA. *Jen.*

PLATE XVI. Fig. 2. Nat. size.

Fig. 2. *b.* Twice nat. size.

Fig. 2. *a.* Magnified scale.

A. gracillima ; corporis altitudine partem vix nonam, capite sextam, longitudinis æquante: oculis mediocribus: maxillis æqualibus, valde protractilibus ; commissurâ primum horizontali, posterius deflexâ : dentibus velutinis, in maxillâ inferiore minutissimis: dorsali primâ omnino pone ventrales reclinatas locatâ : squamis mediocribus, seriebus longitudinalibus duodecim ad maximum dispositis, marginibus liberis inciso-crenatis : vittâ laterali nitidè argenteâ.

D. 5—1/8 ; A. 1/17 ; C. 17 ; P. 12 ; V. 1/5.

D. 5—1/9 ; A. 1/19 ; &c.—

D. 6—1/10 ; A. 1/19 ; &c.—

LONG. unc. 2. lin. 6.

FORM.—Still more slender and elongated than the last species. Greatest depth scarcely one-ninth of the entire length : head one-sixth. Dorsal and ventral lines very little curved. General characters of the head, snout and mouth, as in the *A. microlepidota*, but the jaws more pro-tractile. A row of minute velutine teeth in each jaw most developed above. Eyes moderately large ; their diameter nearly one-third the length of the head ; the space between them just equal to their diameter. Opercle with the posterior margin nearly vertical. Scales larger than in the *A. microlepidota ;* the number of longitudinal rows not exceeding twelve : their form different, and rather peculiar, the anterior or free edge of each scale in some instances pre-senting two or three processes, separated by deep incisions ; in others being irregularly notched or jagged, according to the spot whence taken : the surface is marked with con-centric lines, but there is no fan of striæ on the basal half : the breadth of the scale a little exceeds its length, and the basal margin is irregularly sinuous.

First dorsal answering to the space between the tips of the reclined ventrals and the anal. Length of the second dorsal exceeding the intermediate space between it and the first. From the end of the second dorsal to the caudal is rather more than one-fifth of the entire length. Depth of the caudal fork not exceeding one-third the length of the fin. The anal commences in an exact line with the termination of the first dorsal : rather less than its posterior half answers to the second dorsal. Pectorals rather long, measuring nearly one-sixth of the entire length. Breadth of the silver band one-fourth the depth of the body.

COLOUR.—" Body semitransparent, colourless ; with a bright silver band on each side ; also marked with silvery about the head."—D. The band is remarkably bright, and well defined, much more so than in the last species.

I have ventured to consider this as a new species, though none of the specimens in the collection, amounting to three in number, exceed two inches and a half in length, and are probably not full-sized. The form of the scales is so peculiar, that if it were only the young of some described species, it could hardly fail to be identified by such a character, which is not likely to be affected by age, nor to have escaped the notice of an observer. Yet I can find none answering to it in the " Histoire des Poissons." The silvery band also is remarkably bright ; though the slenderness of the body, another of its peculiarities, is perhaps due to immaturity. The fin-ray formula is somewhat different in the three specimens, as shown above, but in other respects they are similar.

Mr. Darwin's notes state that this species was taken in the month of September, in 39° S. Lat., 61° W. Long., several miles from the land. This last circumstance, indeed, would seem to indicate that the specimens were not so very young, as the fry of most fish keep close in shore.

FAMILY.—MUGILIDÆ.

1. MUGIL LIZA. *Cuv. et Val.* ?

Mugil liza, *Cuv. et Val.* Hist. des Poiss. tom. xi. p. 61.

FORM.—Elongated : the depth contained about five and a half times in the entire length : the head exactly five times: height of the head at the nape two-thirds its own length. Mouth chevron-formed, with a tubercle at the extremity of the lower jaw: lips thin. Some extremely minute teeth in the jaws, but none on the palate or tongue. Suborbital obliquely truncated at the posterior angle, but not dilated towards the extremity; the lower or anterior margin straight, and scarcely if at all denticulated: the maxillary slender, not longer than the suborbital, and concealed beneath it when the mouth is closed. The eye has an adipose veil covering a large portion of the iris: diameter of the orbit one-fourth the length of the head: distance from the eye to the end of the snout, equalling only three-fourths of the diameter. Orifices of the nostrils widely separate. Number of scales in a longitudinal row about thirty-five; perhaps one or two more: in the depth about twelve. Fourth dorsal spine very weak. A large triangular scale above the pectorals; the same also above the ventrals; this last, which is the longer of the two, equalling one-fourth the length of the fin.

D. 4—1/8; A. 3/8; C. 14, &c. P. 16; V. 1/5.

Length 11 inches 3 lines.

COLOUR.—" Back coloured like Labrador felspar: iris coppery."—D. The dried specimen shows traces of about twelve longitudinal lines similar to those of many other species.

A second specimen exactly resembles the above, except in being smaller, measuring barely eight inches, and in shewing rather more trace of denticulations on the suborbital.

Habitat, Bahia Blanca and Monte Video.

This species, which has the general characters of the *M. Cephalus* of the European seas, is probably the *M. liza* of Cuvier and Valenciennes; but the specimens are in a bad state of preservation, and some of the characters cannot be accurately ascertained. The depth of the body appears to have been rather greater than what is mentioned in the " Histoire des Poissons:" there is also some appearance of small scales on the second dorsal and anal, which, according to Cuvier and Valenciennes, is the distinguishing characteristic of their next species, the *M. curema;* but it will not agree with this last in its other details.

The larger of the above specimens was taken at Bahia Blanca, where Mr. Darwin's notes state that it is plentiful; the smaller one at Monte Video.

2. MUGIL ——?

Mr. Darwin's collection contains a second species of this genus from the Keeling Islands, which does not appear to be identical with any of those described by Cuvier and Valenciennes; but as there is but one specimen, in a very bad state of preservation, and the species inhabiting the Indian Ocean are very numerous, as well as extremely similar to each other, I refrain from describing and naming it as certainly new. I shall therefore merely point out some of its leading characters, so far as they can be ascertained; in the hope that they may prove of use in leading others to identify it who may visit the above Islands hereafter.

M

Form and appearance of the mouth similar to that of the *M. labeo* of the Mediterranean. Lips fleshy, and very much developed, with the borders fringed ; the lower one partially reflexed. Apparently no trace of teeth anywhere. Suborbital with a shallow notch on its anterior margin, obliquely truncated at its posterior angle, and obsoletely denticulated. Maxillary slender and slightly bent, nearly concealed beneath the suborbital, but showing a little beneath it, from its being a trifle longer. The head is a little less than one-fifth of the entire length : the snout short, and rather obtuse. Longitudinal diameter of the eye contained three and a-half times in the length of the head : no appearance of any adipose veil. Orifices of the nostril approximating. The depth of the body cannot be accurately ascertained, but it appears to have been about one-fifth of the entire length. The commencement of the anal is but very little in advance of that of the second dorsal ; both fins appear to have been covered with small scales. Pectorals not quite so long as the head : apparently no elongated scale above them : one, however, above the ventrals, half the length of those fins. The fin-ray formula is as follows :—

D. 4—1/8 ; A. 3/9 ; C. 14 ; P. 16 ; V. 1/5.

The length of this fish is eight inches.

DAJAUS DIEMENSIS. *Richards.*

Dajaus Diemensis, *Richards.* in Proceed. of Zool. Soc. 1840, p. 25.

This genus, which was established by Cuvier and Valenciennes, differs from *Mugil* principally in having vomerine and palatine teeth : the snout also is rather more produced, and the mouth less chevron-formed. There is but one species described in the " Histoire des Poisssons," which is found in fresh water in the Caribbee Islands. Dr. Richardson has briefly noticed a second from Van Diemen's Land, in his recent description of a collection of fishes from that country, in the " Proceedings of the Zoological Society." Mr. Darwin's collection contains a specimen of this genus from King George's Sound, which, having reason to think it might be the same as that described by Dr. Richardson, I sent to this latter gentleman, requesting him to compare them. This he obligingly did, and informed me in his answer that he could detect no differences between them, beyond what might be the result of the different manner in which they were preserved, his own specimens being in spirits, and Mr. Darwin's dried.

I forbear giving a detailed description of this species, as one by Dr. Richardson will appear shortly in the Transactions of the Zoological Society ; and Mr. Darwin's specimen is in such a bad state of preservation, as hardly to admit of an accurate description of it being taken. I may just allude, however, to some of its more striking peculiarities.

It appears to differ from the *D. monticola* of Cuvier and Valenciennes in having the teeth in the lower jaw, if they really exist, so minute and thinly scattered as to be scarcely perceptible those in the upper jaw, however, are very distinct ; so likewise are the vomerine and palatine bands. There are also some very obvious teeth on the tip, and at the sides of the tongue, though few in the middle : this part is said to be without any asperities in the *D. monticola*. The suborbital is more rounded off at the lower angle anteriorly, and the denticulations thereon rather more numerous and better developed. The scales on the body, those especially above the lateral line, have a few minute teeth on their free edges, communicating a roughness to the touch ; a character not alluded to in the description of the *D. monticola*, and which therefore may be presumed absent. There are also three more rays in the anal, and one in the second dorsal.

The depth of the body in this specimen, from its bad state of preservation, cannot be ascertained ; but the head is contained about four and a-half times in the entire length. The diameter of the orbit is one-fourth the length of the head ; and there is nearly one diameter between it and the end of the snout. The jaws are nearly equal, but when the mouth is closed, the upper one projects a trifle ; this last is also moderately protractile. The maxillary retires beneath the suborbital. The fin-ray formula is as follows :—

$$D. 4—1/9 ; A. 3/12 ; C. 14, \&c. ; P. 15 ; V. 1/5.$$

There is but one individual of this species in the collection, which measures seven inches in length. The colours do not appear to have been noticed.

FAMILY.—BLENNIDÆ.

BLENNIUS PALMICORNIS. *Cuv. et Val.*

Blennius palmicornis, *Cuv. et Val.* Hist. des Poiss. tom. xi. p. 159.

The Blenny, which I have referred above to the *B. palmicornis* of Cuvier and Valenciennes, seems somewhat intermediate in its characters between that species and the *B. parvicornis* of the same authors. This inclines me to suspect that the two species are not really distinct, as those authors themselves seem to have thought possible, though they state that they never received the *B. palmicornis*, except from the Mediterranean.*

In this specimen the head is one-fifth of the entire length, and the ventrals one-eighth, which is worth noticing, because it is stated that in the *B. palmicornis* the head is contained nearly five and a-half times, and sometimes nearly six times in the total length ; and the ventrals nearly ten times in the same. The filaments above the eyes, however, are similar to those of the species just mentioned ; quite as much developed, and each divided nearly to the base into five or six flattened bristles. There are about forty teeth in the upper jaw, and twenty-eight or thirty in the lower : the canine below is very distinct, but above it is almost, if not quite wanting. The fin-ray formula is as follows :—

$$D. 11/21 ; A. 21 ; C. 11, \&c. ; P. 13 ; V. 2.$$

The length of the specimen is nearly five inches. The anal is marked and coloured exactly as described to be the case in the *B. palmicornis*.

This species was obtained by Mr. Darwin at the Cape Verde Islands.

* According to Mr. Lowe, however, the *B. palmicornis* is common at Madeira, (see *Proc. of Zool. Soc.* 1829, p. 83), and a specimen received from him, undoubtedly belonging to that species, is in the Museum of the Cambridge Philosophical Society.

1. BLENNECHIS FASCIATUS. *Jen.*

PLATE XVII. Fig. 1.

B. flavescens, fusco-variatus; maculis tribus infra pinnam dorsalem, et unâ in pinnæ ipsius anticam partem, nigris, subocellatis: dentibus maxillaribus supra circiter viginti quatuor, subtus triginta; caninis nullis: tentaculis palpebralibus duobus, parvis, subpalmatis: pinnâ anali haud ultrà dorsalem extensâ.

D. 13/16; A. 20; C. 13, &c.; P. 14; V. 2.

LONG. unc. 2. lin. 4.

FORM.—Body much compressed behind : the depth one-fifth of the entire length : head rather less than one-fourth of the same. Snout blunt and truncated ; the profile nearly vertical; the eye placed just within the angle formed by this last with the line of the crown. Diameter of the eye one-fourth the length of the head ; distance between the eyes half a diameter ; the inter-ocular space very slightly concave, with a double row of mucous pores rather widely separate, but without any lines or sculpture. Similar mucous pores are thinly scattered over the occiput and the front of the snout, as well as beneath each eye. Above each eye is a short slightly palmated filament not exceeding in length the diameter of the eye : also an extremely minute one at each nostril. Mouth reaching to beneath the eyes. Teeth not extending the whole length of the jaws ; fine and close-set, with the points of those at the sides, more especially in the lower jaw, reclining backwards ; the number above twenty-four, below thirty : no canines. Gill membrane fastened at bottom, the slit at the sides not descending below the pectorals.

The dorsal commences at the nape, and extends nearly to the caudal, with which, however, it is not connected : it is slightly depressed or notched above the twelfth and thirteenth rays, beyond which it is again elevated to the height of the anterior portion. The anal does not approach quite so near the caudal as the dorsal, but the difference is trifling : the last ray in both fins is united by the membrane to the fleshy part of the tail. Caudal rounded, with the greater part of the principal rays slightly divided at the tips. Pectorals broad, and not quite equal to the head in length. Ventrals short, not more than half the length of the head, or a little less than one-eighth of the entire length : they appear to consist of only two rays, but on dissection there will be found three soft rays with a short spine closely adhering to the first of them ; the third soft ray is slender, and also adheres to the second.

The anterior portion of the lateral line takes a sweep over the pectoral, and is very distinctly marked by a close series of short elevated mucous tubes between two rows of pores ; but the rest of the line is only faintly traced out by nine or ten slender depressed tubes at long intervals, without any accompanying pores.

COLOUR.—(*In spirits.*) Yellowish ground ; the upper half of the sides very much mottled, and clouded with fuscous ; three spots darker than the rest, arranged longitudinally beneath the posterior half of the dorsal, and having a subocellated appearance, the last the largest, and also the most distinct of the three : from the median line there are eight or nine descending fasciæ, alternating with the same number of oblong lanceolate spots : the throat is marked with three angulated transverse dark fasciæ : cheeks and gill-covers with small spots. A large black spot on the first three rays of the dorsal fin, which is covered all over with smaller spots, as are also

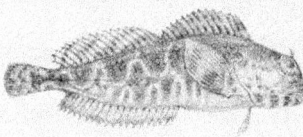

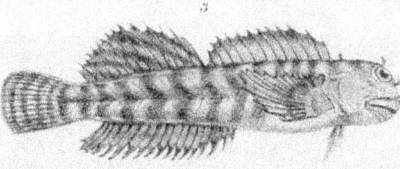

1 *Blennechis fasciatus* Nat. size.
1a , , Teeth magnified
2 *Blennechis ornatus* Nat. size.
3 *Salarias Vomerinus* Nat. size.

FISII.

85

the pectorals and caudal: anal with a dusky edging. In the living state there were probably some bright colours, as in the *B. biocellatus* of Cuvier and Valenciennes.

A second specimen has the fin-ray formula as follows :—

D. 13/18; A. 21, &c.

This specimen also differs from the one above in having the teeth in the lower jaw not quite so numerous, and the ventrals longer, equalling one-seventh of the entire length. The colours are on the whole similar, but more of the yellow ground is visible above the median line, and the descending fasciæ beneath it are not so distinctly traced out.

Habitat, Concepcion, Chile.

This species is very closely allied to the *B. biocellatus* of Cuvier and Valenciennes, from the same coasts. It agrees with it in all its essential characters, and in the general disposition of the markings. It appears to differ, however, in having fewer teeth ; in the anal reaching hardly so far as, certainly not beyond, the dorsal, as described to be the case in that species ; in the fin-ray formula ; and slightly in the colours. The *B. biocellatus* derives its name from two ocellated spots, one *beneath* the last rays of the dorsal, the other *upon* the first three rays of that fin. In the *B. fasciatus* here described, there appear to be three subocellated spots beneath the dorsal, though the last is the most distinct, besides the one upon the fin itself. The *B. biocellatus* was observed by M. Gay at Valparaiso. The present species was taken by Mr. Darwin at Concepcion. Possibly it may be a mere variety.

2. BLENNECHIS ORNATUS. *Jen.*

PLATE XVII. Fig. 2.

B. cinereo-griseus ; maculis, vel lituris paucis, infrà pinnam dorsalem obsoletis, pallidè nigricantibus: dentibus caninis nullis : tentaculis palpebralibus duobus, parvis, subfurcatis : pinnâ anali haud ultrà dorsalem extensâ.

D. 12/17 ; A. 20 ; &c.—

LONG. unc. 2. lin. 2.

FORM.—Closely resembling the last species, but rather deeper in proportion to its length, less compressed in front, with the head more inflated about the throat and gills. Snout, profile, and position of the eye, similar. Superciliary filaments scarcely longer, but rather broader and more conspicuous, and cleft at the extremity. Filaments at the nostrils a little longer, but very slender and delicate. Teeth similar, both in number and form. Fins and lateral line exactly similar. Behind the vent a papilla not present in the last species.

COLOUR.—Different from that of the *B. fasciatus,* but with traces of the same markings. The ground colour is cinereous grey, which almost every where prevails : there are faint traces of the angulated fasciæ beneath the chin, as well as of three dark stains beneath the dorsal, but these last no longer deserve the name of ocellated spots. Fins, cheeks, and gill-covers, dotted in like manner: also some indication of the larger spot on the first three rays of the dorsal: anal with the same dusky edging.

Obs. Of this species there are five specimens in the collection. The next in size to the one described above, measures one inch seven lines in length, and resembles it in every respect, excepting that the superciliary filaments are broader and longer, equalling at least one diameter and a half of the eye. The colours and markings are exactly the same, only the fasciæ on the throat can hardly be discerned.

No. 3 is exactly similar in size, as well as in all its other characters, to No. 2. Has the superciliary filaments equally developed.

No. 4 resembles Nos. 2 and 3, but is smaller, measuring one inch five lines in length.

No. 5, the smallest of all the specimens, and measuring only one inch three lines, has the dark markings more developed, especially the angulated fasciæ on the throat, which are almost as distinct as in the *B. fasciatus :* the spots beneath the dorsal assume the appearance of abbreviated transverse fasciæ reaching from the base of the fin to the median line; and besides the three faintly indicated in the other specimens, there are two others nearer the head, forming altogether a series of five. In this specimen the superciliary filaments are shorter, not exceeding the diameter of the eye.

Habitat, Coquimbo, Chile.

This species differs but slightly from the last, and both may hereafter prove to be mere varieties of the *B. biocellatus*; but it is desirable for the present to keep them distinct, as, though all found on the same coast, they are from distinct localities on that coast. Also the above five specimens, though varying in the intensity of the markings, have all a ground colour quite different from that of the *B. fasciatus*, and a peculiarity of aspect immediately noticeable to the eye. Had they been found mixed with that species, the presence of the anal papilla might lead to the suspicion of their being the other sex ; but, under the circumstances, this seems hardly probable. They were all taken at Coquimbo.

7. SALARIAS ATLANTICUS. *Cuv. et Val.*

Salarias atlanticus, *Cuv. et Val.* Hist. des Poiss. tom. xi. p. 238.

Two individuals of this species were obtained by Mr. Darwin at Porto Praya. They accord in all respects with the descriptions in the "Histoire des Poissons," excepting as regards the fin-ray formula, in which there is a slight difference observable ; and in this respect they are also different from each other.

The larger specimen, measuring three inches seven and a half lines in length, has the fin-ray formula as follows :

D. 13/21 ; A. 24 ; C. 13 ; P. 15 ; V. 2.

The other, two inches eleven lines in length, has one ray less in the spinous portion of the dorsal, and two more in the soft :

D. 12/23 ; A. 24 ; &c.—

It may be mentioned that in this species, as in some others, the last spinous ray in the dorsal is entirely invested by the membrane, and does not attain to the margin, so that in counting, it may be very easily overlooked.

In Mr. Darwin's notes, it is stated that this species bites very severely, having driven its teeth through the finger of one of the officers in the ship's company. Its two very long sharp canine teeth at the back of the lower jaw are well calculated to inflict such a wound.

2. SALARIAS QUADRICORNIS. *Cuv. et Val.?*

Salarias quadricornis, *Cuv. et Val.* Hist. des Puiss. tom. xi. p. 243. pl. 320.

Mr. Darwin's collection contains a species of *Salarias* so closely resembling the *S. quadricornis* of Cuvier and Valenciennes, that I dare not describe it as distinct. Yet it offers some slight differences as follows:

The profile, instead of being merely vertical, presents a rounded and projecting front between the eyes, advancing further than the mouth (as in the *S. gibbifrons*, Cuv. et Val.) The filamentous appendages are similar, but the superciliary ones are shorter than the diameter of the eye: the palmated ones at the nostrils consist of six or seven bristles. The occipital crest is hardly so much elevated; its height being not more than one-sixth or one-seventh the height of the head, and only one-third its own length. The height of the dorsal equals at least half the depth of the body; the depth of the notch above the thirteenth spinous ray is rather more than half its height. The fin-ray formula is—

D. 13/21; A. 25; C. 13, &c.; P. 14; V. 2.

The *colour*, as it appears in spirits, is nearly of a uniform olivaceous brown, with scarce any indication of vertical bands; paler on the abdomen. There are four or five oblique narrow whitish lines on the dorsal, but not very distinct; also two on the anal, more decided: these lines appear to have been bluish, and there are traces of the same colour about the head and gill-covers.

In all other respects it accords exactly with the description in the "Histoire des Poissons," where it is added, in reference to colour, that this species is subject to much variation. Mr. Darwin's specimen measures five inches two lines in length. The number attached to it has been lost, so that there is nothing to shew where it was taken. It is probably, however, from the Keeling Islands, as there is in the collection, from that locality, another specimen, which I have little doubt of being the female of the one above noticed.

This *second specimen* wants the nuchal crest, as is stated to be the case in the female of *S. quadricornis*. It is not full sized, measuring only three inches four lines in length, which may account for the proportions being a little different from those of the adult. The depth is one-sixth of the entire length, or rather less. The filamentous appendages resemble those of the first specimen, but the nasal ones have rather fewer bristles. In the form of the head,

fins, and all its other characters, it is exactly similar. The fin-ray formula is a little
different;

$$D. 13/20 ; A. 23 ; \&c.—$$

The *colours*, also, as they appear in spirits, are rather different. The general ground of
the body is olivaceous grey, but paler than in the male specimen, and inclining to yellowish,
with faint indications of vertical bands, and also a few dark spots towards the tail end.
Dorsal and anal spotted, the former more so than the latter. Mr. Darwin's notes, taken from
the recent fish, merely state,—" with dull red transverse lines."

The *S. quadricornis* is stated by Cuvier and Valenciennes to be very common
at the Mauritius, whence it may not improbably range as far eastward as the
Keeling Islands.

3. SALARIAS VOMERINUS. *Cuv. et Val. ?*

Salarias vomerinus ? *Cuv. et Val.* Hist. des Poiss. tom. xi. p. 258.

PLATE XVII. Fig. 3.

FORM.—Elongated and compressed, the thickest part being in the region of the gills. Greatest
depth contained about six and a-half times in the entire length: thickness at the pectorals
about two-thirds of the depth, or rather more. Length of the head rather exceeding the depth
of the body, and exceeding its own depth by about one-fourth. Snout obtuse; broad and
rounded when viewed from above. Lips crenated at the sides of the mouth, but not in the
middle. Teeth in the jaws moveable, extremely fine and numerous : two long canines at the
bottom of the lower jaw, curving backwards, and fitting into two corresponding holes in the
palate: also a transverse row of minute teeth on the front of the vomer. Profile nearly vertical ; the eyes placed just within the angle formed by it with the line of the crown. Two
broad palmated superciliary filaments, not equal in length to the diameter of the eyes : two
similar ones at the nostrils, each consisting of six or eight bristles : also two short simple
filaments, one on each side of the nape.

The dorsal, which commences a little behind the nuchal filaments, is so deeply notched
behind the twelfth ray as almost to appear like two fins. The height of the anterior or spinous
portion is about two-fifths of the depth : the posterior is more elevated, equalling three-fourths
of the depth : this portion is connected by its membrane with the upper part of the tail, but
does not reach to the caudal, leaving an interval just equal to half the depth of the tail at this
point. The anal commences opposite the eleventh ray of the dorsal, and does not reach so far
as that fin, leaving three times the space between it and the caudal: the first two rays short
and soft, the first scarcely connected by membrane with those that follow ; the membrane
deeply notched between all the rays, excepting the last three, where it is continuous. Caudal
slightly rounded at the extremity. Pectorals broad, but a little pointed when the rays are not
spread out ; longer than the head, the fifth and sixth rays from the bottom being longest. Ventrals short, only half the length of the pectorals, or one-tenth of the entire length, consisting
(which is unusual in this genus) of four distinct rays, two shorter and slender ones, besides the
two ordinary thick ones.

The lateral line is faintly indicated by a fine line which sweeps over the pectorals, and then
passes off straight along the middle. As far as the pectorals reach, the line is continuous :

beyond, it is interrupted, or only marked out by slightly elevated tubal pores at intervals; and it disappears altogether considerably before reaching the caudal.

D. 12/15; A. 18; C. 13, &c.; P. 14; V. 4.

Length 3 inches 2 lines.

COLOUR.—(*In spirits.*) The ground appears to have been pale yellowish-brown: sides marked with numerous approximating dark transverse fasciæ, twelve or fourteen in number: these fasciæ are continued on to the caudal, where there are five, narrower than those on the body. Head marked with black dots and undulating lines; especially two undulating lines commencing on the cheeks behind the eyes, and passing upwards to the nape: upper lip and sides of the throat marked with several fine lines. A row of black dots a little below the base of the anterior part of the dorsal. The fasciæ on the sides extend on to the dorsal, where they take an oblique direction backwards. Anal pale at the base, but with the tips of the rays dusky. Pectorals and ventrals uniformly plain dusky.

Habitat, Porto Praya, Cape Verde Islands.

Cuvier and Valenciennes state that they have received but one species of *Salarias* from the Atlantic Ocean north of the line, the *S. Atlanticus* already noticed. The present is a second found within that range, obtained by Mr. Darwin at Porto Praya. Perhaps it may be a new one; but it is so very nearly allied to the *S. vomerinus* of the above authors, that I consider it hazardous to describe it as distinct. It agrees especially with that species in having vomerine teeth, and four rays in the ventrals, as well as in the general disposition of the markings; but no mention is made in the "Histoire des Poissons" of the nuchal filaments, which, however, may have been overlooked, as they are small and simple, and not very obvious. If it be identical with that species, its range in the Atlantic must be considerable, as the *S. vomerinus* is found on the coast of S. America, near Bahia. Generally speaking the same species are not observed on both sides of that ocean; and perhaps this is an argument for its being distinct: but if so, it is difficult, without the opportunity of a more close comparison, to point out any essential differences by which it may be characterized.

This species appears also to have many points of agreement with the *S. textilis* brought by MM. Quoy and Gaimard from the Island of Ascension; but the colours do not exactly correspond, neither is there any mention made in the description of this last, of the vomerine teeth and four ventral rays, which so peculiarly characterize the one above noticed.

As I feel some doubts with respect to this species being new or not, I have thought it advisable to have it figured, more especially as there is no figure, either of the *S. vomerinus* or *S. textilis*, to both which it is so nearly allied.

N

CLINUS CRINITUS. *Jen.*

PLATE XVIII. Fig. I.

C. fuscus, nigro-maculatus: tentaculis palpebralibus e crinibus octo a radicibus separatis formatis, nasalibus et nuchalibus palmatis, omnibus parvis subæqualibus: pinna anali radiis mollibus viginti quatuor.

B. 6; D. 26/11; A. 2/24; C. 13; P. 13; V. 3.

LONG. unc. 6. lin. 6.

FORM.—Depth one-fifth of the entire length. Head about one-fourth of the same, rather large, with the cheeks and gills a little inflated. Profile falling gently from the nape : the crown scarcely at all convex. Gape reaching to beneath the anterior part of the eye. Lips thick and fleshy, and partly reflexed, much resembling those of a *Labrus*. Lower jaw projecting a little beyond the upper, and inclining upwards to meet it. An outer row of strong conical teeth in each jaw, with a velutine band behind ; the band broad above, but very narrow below. A largish triangular patch of velutine teeth on the vomer, and a smaller one on each palatine. Tongue free and fleshy, smooth. Eyes moderately large, their diameter one-fifth the length of the head ; high in the cheeks, reaching to, but not interrupting, the line of the profile. The superciliary tentacles consist each of eight short bristles, all separate to the root, but forming together a closely compacted series : two on the nape, of the same length as them, are broad and palmated, the upper half only being divided into eight or ten slender filaments : two on the nostrils are similar to those on the nape, only somewhat smaller.

The dorsal commences at the nape, a little behind the nuchal appendages, and has the spinous portion long, and of nearly uniform height, but no where very high. The spines increase very gradually in length as they advance, the first being the shortest : in the middle of the fin, they equal about one-third the depth of the body, or hardly so much : above each is a short filamentous tag, as in the *Labridæ*. The soft portion is nearly twice the height of the spinous. A small interval between the termination of this fin and the caudal. The anal commences under the twelfth spine of the dorsal : its own two spines are very short, and not half the length of the soft rays, which last are not quite so long as those of the dorsal : the membrane between each of the rays is deeply notched. This fin terminates a very little before the dorsal. The caudal, when expanded, appears slightly rounded. Pectorals broad and rounded, about one-fifth of the entire length. Insertion of the ventrals directly underneath the commencement of the dorsal, and both in a vertical line with the posterior margin of the preopercle. These last fins are contained nearly nine times in the entire length.

Body covered with moderately small scales ; the length and breadth of each scale nearly equal, with the basal portion nearly covered by an irregular fan of striæ, eighteen or twenty in number. Head naked, but the crown and upper part of the snout studded with papillæ, terminating upwards in pores. There are rows of minute scales between the rays of the dorsal for about one-third of their height ; also at the base of the caudal and pectorals, but none on the anal. The lateral line commences behind the upper angle of the opercle at one-fourth of the depth ; when opposite the eleventh ray of the dorsal, it begins to bend downwards, and continues falling till opposite the seventeenth ray, when it gets to the middle of the depth ; from that point it passes straight to the caudal.

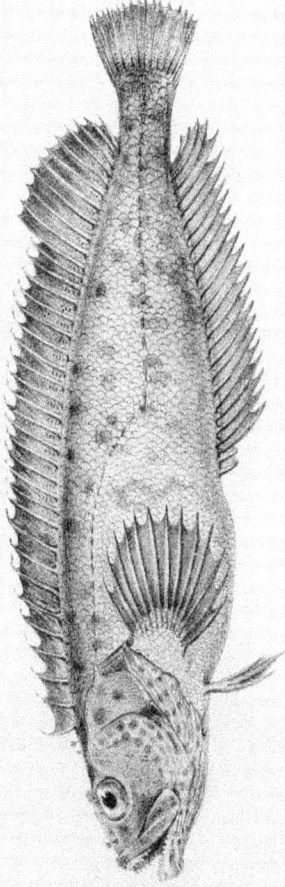

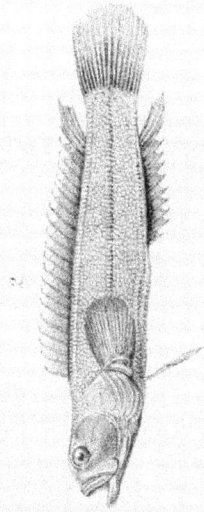

1 Clinus extatus ...... Nat. size.
2 Tripterygius fasciatus Nat. size.

Waterhouse Hawkins del.t

Colour.—(*In spirits.*) Nearly uniform dark brown ground, but with some indications of round black spots, which were probably more conspicuous in the living fish. Eight or nine of these spots appear on the posterior half of the dorsal, forming a longitudinal row ; and there is a row more faintly marked out along the base of the anal ; these last are smaller than those on the dorsal. Chin, throat, and gill-membrane, thickly covered with small spots : also a black patch extending over a large portion of the eye from above and behind.

Habitat, Coquimbo, Chile.

This species, obtained by Mr. Darwin at Coquimbo, is nearly allied to several other Chilian species, described by Cuvier and Valenciennes, but differs from all of them in having more rays in the anal fin, independently of other respects. It seems to approach most closely the *C. variolosus ;* but this latter is represented as having the superciliary tentacles palmated, composed of from twelve to fifteen bristles, and the nuchal ones papilliform and so small as to be hardly visible. In the present species, the superciliary tentacles consist, as above stated, of eight bristles separate quite to the root, while those on the nape are equally as large and as much developed, and strictly, as well as very distinctly, palmated. The crown also is scarcely convex, as represented to be the case in that species : to which it may be added, that the spots on the dorsal fin are more numerous, and their relative size compared with those on the anal different.

The *C. microcirrhis* is said to want superciliary tentacles altogether, otherwise there are several points of resemblance between that species and the one here described.

Genus.—ACANTHOCLINUS. *Jen.*

Corpus elongatum, compressum, squamis minutissimis obtectum. Caput nudum, tentaculis nullis. Dentes maxillares seriebus plurimis dispositi, velutini ; multis, hic illic sparsis, fortioribus, subconicis vel aculeiformibus : vomerini et palatini velutini omnes. Linguæ linea longitudinalis media dentibus minutissimis aspera. Membrana branchialis undique libera, subter gulam continua et profundè emarginata, sex-radiata. Pinnæ dorsalis et analis spinis plurimis, ad apices laciniis membranaceis investitis. Lineæ laterales tres distinctæ.

Mr. Darwin has brought home several specimens of a small fish from New Zealand, which appears to me to form the type of a new genus in the family of the Blennies. It is most nearly allied to *Clinus*, to which group it may perhaps be subordinate in point of value ; but it offers several differences which I shall proceed to point out. In the first place the number of anal spines is much greater, a character of considerable importance in this family, in which they hardly ever amount to more than two, whilst in some instances all the rays of this fin appear to be articulated. Secondly, in addition to the bands of vomerine and palatine

teeth, which are found in *Clinus,* this genus has a narrow line of very minute
teeth running longitudinally down the middle of the tongue, communicating a
sensible roughness to the touch. Thirdly, the ventrals are more backward, their
point of insertion being only a very little in advance of that of the pectorals.
Lastly, it is remarkably characterized by having three, or one might almost
say four, distinct lateral lines. The uppermost of these lines commences
at the posterior angle of the opercle, whence it turns abruptly upwards
and runs immediately beneath the base of the dorsal : the second runs
along the median line of the body, but does not commence till a little beyond
the base of the pectoral : the third commences a little above the insertion of the
ventrals, and answers to the upper one, taking its course a little above the
anal : there is also part of a fourth, which originates between the ventrals, and
joins the third at the commencement of the anal. All these lines are marked by
larger and differently formed scales from those on the body, (which last are very
minute,) with an elevated tube on each, the tubal pore, however, being most
distinct on the middle or second line. In its general form, and in the large
number of *dorsal* spines, this genus resembles *Clinus:* the form of the head and
mouth are for the most part similar; also all the parts of the gill-cover; as well as
the branchial membrane, which is six-rayed and free all round. The tags at the
tips of the dorsal and anal spines are very conspicuous, and give those fins somewhat of a *labriform* appearance.

It is not improbable that the *Clinus littoreus* of Cuvier and Valenciennes, which
they have characterized from a drawing and description in the Banksian Library,
and which is said to possess twenty-five spines in the anal fin, may belong to this
new genus. It is observed by those authors, in reference to its peculiarity in this
respect, that such a circumstance, if correct, would be unexampled, and would
tend to separate it from the genus in which they have placed it. It is also worth
remarking that the *C. littoreus* comes from New Zealand, the same country as
that whence Mr. Darwin obtained the above.

In the circumstance of having three lateral lines, this new genus seems to
have some affinity with *Chirus* of Steller ; but the scales are not ciliated as they
are said to be in this last, neither are the ventrals five-rayed.

ACANTHOCLINUS FUSCUS. *Jen.*

PLATE XVIII. Fig. 2.

FORM.—Body elongated and compressed ; the depth, which varies but little, one-sixth of the entire
length ; thickness in the region of the pectorals rather more than half the depth. Head contained very little more than four times in the length. Profile sloping but very little. Snout
rather short : mouth protractile, and rather wide : lips somewhat fleshy and reflexed. Gape
reaching to beneath the anterior part of the orbit, but the maxillary, which is dilated at its

posterior extremity, and cut nearly square, reaching to beyond the middle. Lower jaw a little the longest, and ascending to meet the upper. Several rows of sharp velutine teeth in each jaw, with some here and there stronger and more hooked than the others, those below almost fine card : a band on the vomer and on each palatine. Tongue of a triangular form, free and pointed at the tip, with a ridge of asperities down the median line. Eyes high, but hardly interrupting the line of the profile ; their diameter one-fifth the length of the head ; distant one diameter from the end of the snout. No filamentous appendages of any kind on any part of the head ; but an irregular circle of pores nearly surrounding the orbit ; also a few very distinct pores beneath the lower jaw. Preopercle rounded, with distant pores along the margin. Opercle terminating posteriorly in a sharp salient angle with the basal margin ascending ; beneath which the subopercle and interopercle are both very distinct. Branchial membrane free and open all round, not adhering to the isthmus underneath, but deeply notched in the middle.

The dorsal commences in a line with the posterior point of the gill-cover, and is very similar to that of *Clinus.* Spinous portion long, and, excepting the first two rays, of nearly uniform height, equalling nearly half the depth ; the membrane deeply notched between the spines, the tips of which are invested with filamentous tags. Soft portion of the dorsal more elevated than the spinous, and with only four rays. Between the end of this fin and the caudal is a small space equalling nearly two-thirds of the depth beneath. The anal commences under the twelfth dorsal spine, and exactly corresponds to the posterior half of that fin, reaching also to the same point. The spines in both fins are sharp and moderately strong ; the soft rays articulated and branched, and terminating rather in a point behind. Caudal rounded, with fourteen branched rays, and a few shorter simpler ones. Pectorals one-seventh of the entire length, rounded when spread open, with all the rays except the last branched. Ventrals narrow and pointed, about the same length as the pectorals, and inserted but very little in advance of those fins : the spine well developed, and half the length of the soft rays : first soft ray long, and deeply divided so as to appear like two ; the second ray slender and shorter.

Body covered with very minute scales ; but none on the head or on any of the fins. Three very distinct lateral lines, with a portion of a fourth, as already stated above.

B. 6 ; D. 20/4 ; A. 9/4 ; C. 16, &c. ; P. 17 ; V. 1/2.

Length 3 inc. 8 lin.

COLOUR.—Not noticed in the recent state. *In spirits* it appears of a nearly uniform bister brown, with the fins and some portion of the head darker than the rest, especially a blackish spot on the opercle.

Habitat, Bay of Islands, New Zealand.

There are four specimens of this new fish in the collection, all similar except in size. The above is the largest. The others measure in length from one inch and three quarters, to not quite three inches. The two largest are from the Bay of Islands, New Zealand. The other two have lost their labels : I only presume therefore that they are from the same locality.

Tripterygion Capito. *Jen.*

Plate XIX. fig. 1.

T. fusco-griseum, pinnis concoloribus: tentaculis palpebralibus duobus parvis gracilibus e crinibus duobus vel tribus formatis ; nasalibus minutis simplicibus: dorsali primâ humili sex-radiatâ, radiis subæqualibus ; secundâ duplo altiore ; tertiâ purum altissimâ : lineâ laterali abbreviatâ, vix ultrà pectorales extensâ.

B. 6; D. 6—20—14; A. 25; C. 14, &c. ; P. 16; V. 2.

Long. unc. 2. lin. 5.

Form.—Depth at the pectorals one-sixth of the length : thickness at the same part about two-thirds of the depth. Head rather large, thicker than the body, contained four and a half times in the entire length. Snout short, the profile falling very abruptly from between the eyes. These last large, one-third the length of the head, high in the cheeks, reaching to, but hardly interrupting, the line of the profile. Above each a short slender compound tentacle : that on the right side consists of two filaments, one simple, the other forked, so as to appear like three ; that on the left appears undivided. Also a minute filament at each nostril. The maxillary reaches to beneath the middle of the orbit. Jaws equal : in each a row of small conical sharp-pointed teeth, with a broad velutine band behind, the band, however, only in front. A transverse band of velutine teeth on the vomer, extending a little on to the palatines. Opercle and preopercle rounded. Branchial membrane free all round, with a shallow notch in the middle underneath.

The first dorsal commences in a vertical line with the insertions of the ventrals ; the rays are six in number, and so nearly equal in length as to cause the fin to appear quite even ; its height is scarcely more than one-third of the depth. The second dorsal begins a little behind the origin of the pectorals : it is also nearly even, but twice the height of the first. The third closely follows the second : this fin is uneven, but its most elevated point is somewhat higher still than the.second. The rays of the first and second of these fins are spinous : those of the third soft and articulated, but all simple. The anal, which has also simple rays, commences beneath the middle of the second dorsal, and terminates in the same vertical line with the end of the third, between which last and the caudal is a small space. Caudal square, with twelve of the principal rays branched. Pectorals a little less than one-fourth of the entire length ; the ninth and tenth rays longest ; the six lowermost rather stouter than the others, and, as well as the three uppermost, which are very slender, simple ; the fourth to the tenth, both inclusive, branched. Ventrals contained about six and a half times in the entire length ; consisting of only two slender filamentous rays.

Scales minute, their free edges finely ciliated ; the concealed portion of each scale marked with twelve or fourteen striæ. The lateral line rises at the upper angle of the opercle, and is well marked by a row of tubular scales till it reaches a little beyond the extremity of the reclined pectoral, where it abruptly terminates, and all further trace of it is lost.

Colour.—(*In spirits.*) Of a nearly uniform dark brown, inclining to griseous, with some appearance of darker clouds or spots between the second dorsal and the lateral line; this last also is indicated by a darker streak than the ground colour. Fins dark brown : there is, however, some trace of a white edging to the anterior half of the anal, which may have been more conspicuous in the living state.

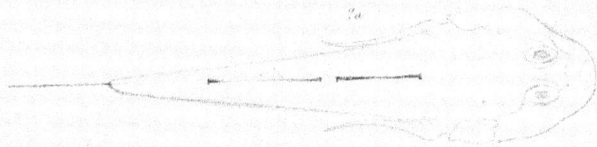

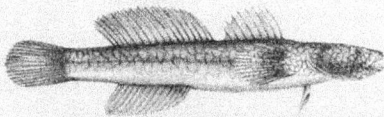

Waterhouse Hawkins del.

1 Tripterygion Capito.
2 Gobius lineatus.
2a   dorsal view
3 Gobius epicephalus
3a   dorsal view

A *second specimen* slightly differs from the above, but is evidently referable to the same species. It is smaller; and the profile falls more gradually. The caudal has only eight branched rays, with two lateral simple ones. The pectorals have the tenth and eleventh rays longest, with the seven lowermost (instead of six) stouter than the others and simple. The fin-ray formula is also different.

D. 6—19—13; A. 25; C. 10, &c.; P. 17; V. 2.

Length 2 inc. 1 line.

The colours are paler, and more decidedly grey, with the darker motlings more distinct. The dorsals and caudal are pale, minutely dotted with brown. Tips of all the anal rays white.

Habitat, Bay of Islands, New Zealand.

This species approaches very closely the *T. nigripenne* of Cuvier and Valen-ciennes, of which it may possibly be a variety; but the description in the "Histoire des Poissons," as regards the form, is limited to a very few words. If the figure given by those authors be correct, the *T. nigripenne* differs decidedly in the first dorsal being more elevated, with the rays more unequal, and in the lateral line extending the whole length of the fish. In the present species the first dorsal is low and even, with the rays all equal, and the lateral line cannot be traced much beyond the pectoral; and these characters are found in both specimens. There are also six rays in the first dorsal. According to the description, the *T. nigripenne* has but five, though six are represented in the figure.

From the *T. varium*, this species differs not only in its fin-ray formula, but in its markings: and the same characters serve to separate it still more widely from *T. Forsteri* and *T. fenestratum*.

This species was obtained by Mr. Darwin on tidal rocks in the Bay of Islands, New Zealand. Three out of the only four extra-european species described by Cuvier and Valenciennes come from the same locality.

FAMILY.—GOBIDÆ.

1. GOBIUS LINEATUS. *Jen.*

PLATE XIX. fig. 2.

G. nigro-grisens, lineis circiter decem longitudinalibus nigris: capite lato, subdepresso; genis inflatis: maxillis æqualibus: dentibus velutinis, externis fortioribus aculeifor-mibus; caninis nullis: oculis amplis, intervallo vix plus quam semidiametrum æquante: pinnis dorsalibus contiguis, altitudine subæqualibus; pectoralibus radiis supernis setaceis, liberis; caudali rotundatá: squamis mediocribus, levissimè ciliatis.

B. 5; D. 6—1/9; A. 1/8; C. 13, &c.; P. 7 et 16; V. 1/5.

LONG. unc. 4. lin. 8.

FORM.—Head large, sub-depressed, and much inflated about the gills : body compressed towards
the tail. Depth at the pectorals contained about five and a half times in the length : thickness
at the same point about three-fourths of the depth. Head about four and a half times in the
length ; its breadth nearly equal to its own length. Profile nearly horizontal. Eyes moderately
large, with a diameter nearly one-fourth that of the head : the intermediate space a little hollowed
out, and scarcely more than half a diameter in breadth. Some appearance of a shallow groove
on the nape reaching to the first dorsal. Gape reaching to beneath the anterior angle of the eye.
Jaws equal : each with a broad band of velutine teeth, the outer row stronger than the others,
and slightly hooked ; of these stronger ones there are twenty six in the upper jaw ; below they
are fewer, smaller, and more irregular : no canines : no vomerine or palatine teeth.

Pectorals about one-fifth of the entire length, oval ; the first six or seven rays nearly free
to their base, and setaceous, like those of G. niger ; the sixteen that follow connected by membrane as usual, and much branched. Ventrals united in the usual manner, and a little shorter
than the pectorals. The first dorsal commencing a very little behind the point of attachment of
the pectorals, and reaching to the extremity of those fins when laid back : the anterior spines
rather exceeding in length half the depth of the body ; the last three gradually decreasing,
with the membrane terminating at the foot of the second dorsal. This last fin with the first ray
simple, and of the same height with the anterior rays of the first dorsal ; those which follow, to
the number of nine, nearly of the same height, and branched ; from the root of the ninth springs
a simple ray which might be reckoned as distinct, and if so, the entire number would be ten.
Anal commencing a little more backward, and terminating a little sooner than the second dorsal,
to which in other respects it answers ; the last ray double as before : both these fins terminate
in a point behind. Space between the anal and the caudal rather more than one-fifth of the
entire length, and equalling twice the depth immediately beneath. Caudal rounded, about
one-sixth of the entire length ; the division between the principal and accessory rays (which last
are numerous, especially above), not well marked ; the former much branched. The usual
papilla behind the vent.

No visible lateral line. Scales rather large; about thirty-seven in a longitudinal line,
and eleven in a vertical ; ciliated, the concealed portion of each scale with an irregular fan of
very numerous striæ, amounting to twenty-five or more. Skin of the suborbital marked with
four longitudinal lines of salient dots, the third from the top forking posteriorly into two : a
similar line at the upper part of the opercle at the boundary of the scales, whence another passes
vertically across the branchial membrane ; behind this is a third shorter one, taking an oblique
direction backwards.

COLOUR—(In spirits.) Dusky grey, with about ten, rather indistinct longitudinal dark lines on the
body, extending from the pectorals to the caudal. Fins dusky, with some indication of small
irregular whitish spots scattered here and there. A dark spot on the upper half of the eye.

Habitat, Galapagos Archipelago.

This is undoubtedly a new species. It belongs to the same section as the
G. niger of the European seas, which in form it very much resembles, especially
in its large inflated head, and in having the uppermost rays of the pectorals free
and setaceous. It differs, however, in having fewer rays in the dorsal and anal

fins, and consequently a larger interval between the anal and the caudal ; also, in the number and arrangement of the dotted lines on the cheeks. The colours are likewise different ; and, in the living fish, in which they were not noticed, probably the dark longitudinal lines, alluded to in the description above, are much more conspicuous than they are at present.

This species was taken by Mr. Darwin off Chatham Island, in the Galapagos Archipelago.

2. GOBIUS OPHICEPHALUS. *Jen.*
PLATE XIX. FIG. 3

G. pallenti-plumbeus, fusco-reticulatus : corpore elongato, gracili, undique alepidoto : capite lato, depresso, genis tumidis ; his et rostro punctis valde salientibus, creberrimis, lineis undantibus dispositis: maxillis æqualibus : dentibus velutinis ; externis, præsertim lateralibus, fortioribus, aculeiformibus ; caninis nullis : oculis parvis, prominulis, intervallo plus quam diametrum æquante : pinnis dorsalibus subcontiguis, altitudine subæqualibus ; pectoralibus radiis omnibus membraná inclusis; caudali rotundatá, radiis clausis, subacutá.

D. 8—1/16 ; A. 1/13 ; C. 17, &c. ; P. 21 ; V. 1/5.

LONG. unc. 2. lin. 11.

FORM.—Body considerably elongated, and compressed posteriorly: the greatest depth beneath the first dorsal, equalling rather less than one-eighth of the entire length : thickness at that point rather less than the depth. Head broader than the body, very much flattened in the crown behind the eyes, with the cheeks tumid, and, on the whole, snake-like in appearance : its length one-fifth of the entire length ; its breadth two-thirds of its own length. Eyes small, but rather prominent, high in the cheeks, with a diameter scarcely exceeding a line in length, or about one-sixth that of the head ; the space between a little hollowed out, and nearly a diameter and a half across. Snout short and obtuse : jaws equal; the gape not quite reaching to beneath the middle of the orbit. The teeth form a broad velutine band in each jaw, with those in the outer row strong and slightly hooked : of these last there are about twenty in the upper, the lateral ones being stronger than those in front ; in the lower they are not so numerous, and more irregular : none that can be strictly called canines : likewise no vomerine or palatine teeth.

Pectorals one-sixth of the entire length, oval, with the middle rays longest; all the rays included in the membrane. Ventrals united; about two-thirds the length of the pectorals. First dorsal extending beyond the extremities of the pectorals ; the rays very gradually decreasing in length, the membrane beyond the last also sloping very gradually down till it nearly reaches the second dorsal, which it does not quite touch. Rays of the second dorsal of nearly uniform height, about equalling the longest of those in the first, also equalling the depth of the body beneath. The last ray in both these fins is double, as in the last species. The anal commences beneath the fourth ray of the second dorsal, and terminates a little sooner than that fin. The caudal, when the rays are spread, appears rounded ; but when closed, somewhat pointed :

it is contained not quite six and-a-half times in the entire length. The space between the anal and the caudal is one-eighth of the same, and one and a half times the depth of the tail at that part. The usual papilla appears behind the vent.

Skin apparently quite naked everywhere, and without any scales that are visible, even in the dried state, under a lens. The lateral line runs straight along the middle, and is marked by a series of glandular dots placed in threes or fours together vertically at moderate intervals. Several lines of dots about the head, but the dots are here closer together, and in some places so salient as to appear like short filamentous processes : on the cheeks, about the eyes, and on the front of the snout, these lines undulate in an irregular manner: there are also two or three short lines of dots on the gill-cover, and a double row on each side of the lower jaw, passing obliquely upwards posteriorly, as a boundary to the cheek.

Colour.—" Pale lead-colour, coarsely reticulated with brown."—D.—This is nearly as it appears also in spirits. The reticulations are finer on the head, where they are also most distinct: they are likewise very visible at the base of the pectorals.

Habitat, Chonos Archipelago, South of Chiloe.

Cuvier and Valenciennes seem to have doubted * whether there were really any species in this genus absolutely without scales, though they have established a section, in which the scales are very minute, and as it were lost in the skin. The present one, however, appears to be thus characterized : at least there are no scales which can be detected, even with the assistance of a lens, and when the skin is suffered to become dry, in which state they are generally visible, if really present. In fact, the skin is as smooth and naked as in any of the true Blennies. This character, combined with others, clearly indicates it to be a new species ; neither will it assimilate with any of the sections in the " Histoire des Poissons ;" but requires to be placed in one by itself, in which the absence of scales is coupled with an elongated body, and a caudal, not strictly pointed, but approaching to that form, when the rays are close.

This species was obtained by Mr. Darwin in the Chonos Archipelago, in Lowe's Harbour, S. of Chiloe. It appears to be the first of this genus brought from the West Coast of America ; at least, there are none, amongst the very numerous species described by Cuvier and Valenciennes, which are mentioned as belonging to those shores.

ELEOTRIS GOBIOIDES. *Cuv. et Val.*

Eleotris gobioides, *Cuv. et Val.* Hist. des Poiss. tom. xii. p. 186.

This species was taken by Mr. Darwin in fresh-water, in the Bay of Islands, New Zealand. It so well accords with the description of the *E. gobioides* in the " Histoire des Poissons," that I conceive there can be no doubt of their identity.

* See " Hist. des Poiss." tom. xii. p. 72, under the species *Gobius Boscii.*

The profile slopes very gently. The lower jaw is longest, ascending to meet the upper. There are three or four longitudinal lines on the sides of the head, especially a very well marked one (not particularly noticed by Valenciennes) extending backwards from the posterior angle of the eye to the upper angle of the gill-opening. No appearance of any lateral line. This specimen has a ray more in the anal than Valenciennes gives. The fin-ray formula is as follows:—

B. 6; D. 6—1/10, the last double; A. 1/10, the last double; C. 16, &c.;
P. 18; V. 1/5.

Length 4 inches 1 line.

This species, except in respect of its separate ventrals, has very much the habit and general appearance of the *Gobius niger* of the European seas.

FAMILY.—LOPHIDÆ.

BATRACHUS POROSISSIMUS. *Cuv. et Val.?*

Batrachus porosissimus, *Cuv. et Val.* Hist. des Poiss. tom. xii. p. 373.

FORM.—Head very large, broad and depressed, exactly one-fourth of the entire length; its breadth two-thirds of its own length. Body compressed posteriorly, with its greatest depth about one-sixth of the entire length. Snout blunt and rounded, the lower jaw projecting; gape wide. The teeth above form but a single row along the intermaxillary, mostly small, but sharp, and the posterior ones much curved: along each palatine there is a row of much stronger ones, and at each angle of the vomer are two very long hooked ones, resembling true canines. In the lower jaw the teeth are in a single row at the sides, but in two or three rows in front, and are unequally sized, some of the lateral ones being as strong as those on each side of the vomer, and much hooked, as well as partially reclining backwards. Tongue smooth, and free at the tip, which is bluntish. Pharynx armed with two patches of velutine teeth above and below. No regular barbule at the chin, but a row of minute cutaneous cirri running all round the edge of the lower jaw; a similar row along the anterior edge of the upper jaw, behind the intermaxillary, with two thicker and more conspicuous appendages of the skin in the middle. Eyes far apart, and not very large. Opercle armed with one very strong spine, but only just the point appearing through the skin.

Two small spines in front of the dorsal, a little more backward than the insertion of the pectorals, the first very minute, and hardly appearing through the skin. Second or true dorsal very long, reaching to the base of the caudal, and of nearly uniform height throughout, equalling about one-third of the greatest depth of the body; the rays branched, and the membrane notched between their tips. Anal commencing under the fifth dorsal ray, similar to that fin, but with the membrane more notched between the rays: both fins are fastened down at their extremities to the fleshy part of the tail by a membrane. Caudal slightly rounded, when spread. Pectorals broad and large, but, from the middle rays being longest, appearing somewhat wedge-shaped, not quite equalling the length of the head. Ventrals much smaller, only half their length, and cut nearly square.

Skin perfectly naked.   The lines of pores, which are very numerous about the head and body, run in the exact directions laid down by Cuvier and Valenciennes, in their description of the *B. porosissimus;* but in addition to those which have been pointed out by them, there is one commencing at the nostrils, and passing underneath each eye, thence ascending a little behind the eye to descend again by the margin of the preopercle ; another directed transversely across the cheek, connecting the former with the row that passes along the edge of the lower jaw : this transverse row, if continued upwards, would form a tangent to the posterior part of the orbit.   All the lines of pores are furnished with very minute cutaneous appendages, similar to those already spoken of above, as fringing the edges of the jaws.

D. 2—36 ; A. 33 ; C. 12, &c. ; P. 20 : V. 1/2.

Length 9 inches.

COLOUR.—" Above purple-coppery ; sides pearly ; beneath yellowish, with silver dots in regular figures ; iris coppery."—D.   The silver dots alluded to by Mr. Darwin, are the lines of pores. There are two longitudinal dark lines on the dorsal, the uppermost serving as an edging : the anal also is edged in the same manner, especially posteriorly.

Habitat, Bahia Blanca.

This species was found by Mr. Darwin cast up on the beach at Bahia Blanca, where he states that it is not uncommon.   It approaches so closely the *B. porosissimus* of Cuvier and Valenciennes, that I dare not consider it as distinct without comparison.   Yet it differs from their description of that species, in having four vomerine teeth, instead of two ; in having six more rays in the anal fin ; and in having the additional lines of pores above indicated ; though these last may have been accidentally left unnoticed.   It requires the examination of more specimens to determine whether these differences result from a difference in species or not.

FAMILY.—LABRIDÆ.

COSSYPHUS DARWINI. *Jen.*

PLATE XX.

C. corpore elongato-ovali ; capite grandi, fronte elevato, rostro ex hoc declivi : caninis quatuor fortibus ad apicem utriusque maxillæ, ad angulos oris nullis ; dentibus lateralibus conicis ; interiùs, ad latera palati, granis plurimis minutis obtusis : preoperculo, limbo excepto, operculo, et interoperculo, squamatis ; preoperculo margine integro : rostro, maxillis, et suborbitalibus ante oculos, nudis : lineâ laterali subrectâ : pinnâ dorsali parte spinosâ humili, spinis ad apices laciniatis ; molli, heic respondente anali, duplò altiore, sub-acuminatâ : caudali æquali, solùm radiis exterioribus aliis paulo longioribus.

D. 12/10 ; A. 3/12 ; C. 14, &c. ; P. 17 ; V. 1/5.

LONG. unc. 19.

Glyphidus Darwinii R. m. de.

FORM.—Head large : body of a suboval form, but much elongated : greatest depth at the nape contained about four times and three quarters in the entire length : head not quite three times and three quarters in the same. Nape and forehead high, whence the profile descends obliquely in a straight line to the end of the snout. Jaws equal, and rather acute : lips fleshy : the end of the maxillary not quite reaching to a vertical line from the anterior margin of the orbit. Four very conspicuous, strong, curved, canine teeth at the anterior extremity of each jaw; those above of nearly equal length, but the two middle ones rather longer and stouter than the other two; of those below, on the contrary, the outer ones are the longest, as well as strongest, being nearly twice as much developed as the middle ones, which last are of about the same length as, but rather slenderer than, the outer ones above. The teeth at the sides of the jaw are short and conical, and not very sharp pointed, forming a regular series; below they amount to nine or ten on each side; above, the series may have been originally of the same number, but in this specimen several appear wanting. Besides these conical teeth at the sides of the jaws, there is an inner band of small rounded grains about the size of pins' heads: the band is broader, and the grains larger and more distinct above than below : many of them appear much flattened, and as if ground down by use. Eyes of moderate size; their diameter about one-seventh the length of the head; rather high in the cheeks, and nearly equidistant from the end of the snout and the posterior angle of the opercle. Snout and suborbital in advance of the eyes, as well as the jaws, naked. Preopercle large; occupying the posterior half of the cheek, rectangular, but the angle at bottom much rounded, the ascending margin vertical, both margins entire; covered with small scales; the limb rather broad, bounded internally by a slightly raised ridge, and without scales, but with a few scattered small pores. The opercle and subopercle form together an irregular oblong, of which the height is double the length; both are covered with scales larger than those on the preopercle : the membrane terminates behind in a blunt angle. The interopercle, which is very distinct, has three rows of scales on its surface, but none on the margin.

The lateral line is nearly straight throughout its course, the bend downwards beneath the termination of the dorsal fin being scarcely perceptible. The tubes of which it is composed are unbranched; many of them, however, incline upwards at their posterior extremity towards the back. The scales on the body are rather larger than those on the opercle : there appear to be upwards of fifty in a longitudinal line. The free portion of each scale has its surface finely granulated in the middle, and striated at the sides.

The dorsal commences rather before one-third of the entire length, excluding caudal, and occupies a space equalling nearly half the same; the spinous portion is low, and the spines of nearly the same length, the first and second only being rather shorter than the succeeding ones; the membrane between the spines notched : the soft portion rather pointed, and twice as much elevated as the spinous. The anal commences beneath the eleventh or twelfth dorsal spine, and terminates in the same vertical line with that fin; the soft portion, which answers to the soft portion of the dorsal, is preceded by three spines, increasing in length to the third, which is double the first, though itself not above half the length of the soft rays; these spines are not particularly stout. The space between the anal and caudal equals one-sixth of the whole length. Caudal rays nearly even, with the exception of the two outermost above and below, which being rather longer than the others, give the fin a slightly crescent-shaped form : the base of the caudal is scaly, but the scales advance only a very little way between the rays. Pectorals very little in advance of the ventrals, in length more than half that of the head, with the second, third, and fourth rays longest. Ventrals in an exact vertical line with the commencement of the dorsal, nearly equal to the pectorals, with the first and second soft rays longest; the spine rather more than half the length of the first soft ray; the last soft ray united to the body by a membrane.

COLOURS.—" Centre of each scale pale vermilion red: lower jaw quite white: a large irregular patch above the pectoral bright yellow: iris red, pupil blue-black."—D.   The dried skin in its present state is of a nearly uniform brown.

Habitat, Chatham Island, Galapagos Archipelago.

I have named this species in honour of Mr. Darwin, whose researches in the Galapagos Archipelago, where he obtained it, have been so productive in bringing to light new forms.   I have referred it to the genus *Cossyphus* of Valenciennes, on account of the small rounded grains behind the principal teeth ; but it rather departs from that group in not having the preopercle denticulated, and in having no scales on any of the vertical fins, with the exception of a few at the base of the caudal.   In some respects it seems intermediate between that genus and *Labrus*. It does not appear to be described, though it seems to approach the *C. reticulatus* of Valenciennes in many of its characters.   That species however is from Japan.

The canines at the anterior extremity of each jaw are very conspicuous in this fish, and give it at first sight much the appearance of a *Dentex*.

CHEILIO RAMOSUS. *Jen.*

C. nigro-fuscus, infrà lineam lateralem et in ventre obscurè argenteus ; pinnis pallidè fuscis immaculatis : corpore valde elongato : dentibus in maxillâ superiore duobus anticis caninis fortibus, lateralibus conicis parvis subæqualibus ; in inferiore, caninis parvis, lateralibus variis inæqualibus : lineâ laterali ramosâ.

B. 6 ; D. 9/13 ; A. 3,12 ; C. 12, et 4 breviores ; P. 11 ; V. 1/5.

LONG. unc. 9. lin. 6.

FORM.—Very much elongated, with the dorsal and ventral lines nearly straight. Depth varying but little, and contained nine and a half times in the entire length; thickness not quite three-fourths of the depth.   Head elongated, contained not more than three and a half times in the entire length, compressed, with the cheeks vertical.   Snout very much produced, slightly rounded at the extremity: gape reaching half way to beneath the middle of the eye.   Jaws scarcely protractile ; the upper one a little the longest: lips reflexed in the form of membranaceous flaps, especially the lower one, the margin of which is sinuous.   Teeth ranged in a single row in each jaw.   Those above form a numerous, close-set, nearly even series at the sides of the jaw, with two long hooked canines in front ; the lateral teeth amount to about thirty-five on each side, and are small, but strong, somewhat conical, and not very sharp-pointed.   In the lower jaw there are two front canines, similar to those in the upper, but much smaller ; then follow four short conical teeth ; then six large triangular, compressed, sharp-pointed ones, but not all of equal size ; then five more small conical ones, which complete the series on each side. No teeth on the vomer or palatines. Eyes of moderate size, situate in the middle of the length of the head, high, but not touching the line of the profile ; their diameter one-eighth the length of the head. Preopercle rectangular. Opercle triangular, the membrane produced posteriorly at the upper part in the form of a rounded angle : a short row of scales observable along its upper margin, and another along its lower. Subopercle and interopercle without scales. Also a short row of scales, similar to those on the opercle, behind and partially beneath each eye, and, with these exceptions, no other scales on the head. Above each eye is an irregular row of minute pores: there are also pores beneath the eye, and on the sides of the snout, mixed with short raised lines having somewhat the appearance of written characters. Gill-opening widely cleft; the branchial membrane free all round.

Scales on the body moderately large, and similar in form to those of the *C. auratus*, as described by Cuvier and Valenciennes. The number, in a longitudinal line from the gill to the caudal, is forty-six, in a vertical about seventeen. Lateral line also as in that species, but with the mucous tubes branched, and giving off eight or nine twigs on each side.

The dorsal commences a little behind the terminating angle of the opercle, and the anal immediately beneath the first branched ray of the dorsal : these fins terminate in the same vertical line, and the last ray in each is double: the simple rays are soft and flexible. Caudal slightly rounded. Pectorals short, and obliquely truncated, contained eleven and a half times in the entire length. Ventrals very small, about two-thirds the length of the pectorals, rounded, close together, with an elongated scale between them ; their point of insertion slightly backwarder than that of the pectorals.

COLOUR.—Not noticed in the recent state. In spirits, it appears of an almost uniform dark brown, at least above the lateral line. There is some trace of a pale longitudinal band on each side of the head beneath the eye, which is continued, but rather indistinctly, along the whole length of the body, the tips of the scales remaining dark. Possibly during life all the lower part of the sides and belly may have exhibited numerous dark spots upon a pale or silvery ground. Under part of the head pale brown, with some faintly-defined ocellated spots : also a faint trace of red on the opercle. All the fins pale brown, without spots.

Habitat, Japan ?

This species was given to Mr. Darwin, when at Chiloe, by the surgeon of a whaling-ship, who said that he believed that it was caught in the Japan seas. From the great similarity which prevails amongst the species of this genus, I am not sure that it is really new, as I have ventured to consider it. The specific character also, so far as the colours are concerned, must be received with some caution, in consequence of these last not having been observed in the recent state. It seems to approach very closely the *C. hemichrysos* of Cuvier and Valenciennes, brought by MM. Quoy and Gaimard from the Sandwich Islands ; but it differs in its colours, especially in the fins being all uniformly pale brown, and in having fewer scales on the opercle, and beneath the eye. It is impossible to say, however, to what extent the colours may have been altered by the spirit: some of the scales also may have been rubbed off.

CHROMIS FACETUS. *Jen.*

C. supra virescenti-niger, lateribus pallidioribus : dorso modice arcuato ; fronte elevato, rostro summo ante oculos paululum excavato : limbo preoperculi poris quatuor conspicuis impresso : squamis latis, marginibus liberis levissime ciliatis: spinis dorsalibus quindecim, analibus sex: pinnis ventralibus longe acuminatis, ad analem pertingentibus: pinnâ caudali subæquali.

D. 15/10; A. 6,8; C. 16, &c.; P. 14; V. 1/5.

LONG. unc. 5. liu. 9.

FORM.—Oblong-oval, very much compressed ; the back moderately elevated, and more curved than the abdomen. Greatest depth a little behind the insertion of the pectorals, and contained twice and three-quarters in the entire length: thickness about two-fifths of the depth. Forehead high : profile falling very obliquely, and slightly hollowed out in front of the eyes ; the upper and under profile meeting at the mouth at nearly a right angle. Head contained not quite four times in the entire length ; its own length and height nearly equal. Mouth small, protractile : jaws about equal, the lower one, if anything, a little the longest: lips not very fleshy. Maxillary rather slender, retiring almost entirely, when the mouth is closed, beneath the sub-orbital, the anterior margin of which is slightly hollowed out, and somewhat sinuous. Teeth in card in both jaws, forming a narrow band ; the outermost row longer and stronger than the others, especially the four or six middle ones in front, which are somewhat conical and slightly hooked. Pharyngeal teeth present, but none on the vomeror palatines. Eyes rather small, their diameter about one-fifth the length of the head ; high in the cheeks, and a little nearer to the snout than to the posterior margin of the opercle : the space between broad, equalling nearly two diameters and-a-half. Nostrils consisting of a single round orifice half-way between the eye and the end of the snout. Preopercle with the basal margin short, and forming a slightly obtuse angle with the ascending one, the margin of which is entire. Opercle of a triangular form, broad at top, but narrowing off towards the bottom. Subopercle and interopercle much developed ; their outer margins, taken together, rounded off nearly in a semicircle. Branchial membrane quite free all round, unattached to the isthmus, and but slightly emarginate. Snout, suborbital, jaws, and limb of the preopercle, naked ; but the cheeks and rest of the opercular pieces scaly: the scales on the subopercle large. Four large pores on the limb of the preopercle, preceded by three others beneath the lower jaw : similar pores beneath the eye, and extending partially round it ; one on the crown of the head, and a few smaller ones scattered about the snout ; a large one just above the opercle, and another higher up on each side of the nape.

Scales on the body large ; about twenty-five or twenty-six in a longitudinal row, and eleven or twelve in the depth ; broader than long, with the free edges very minutely ciliated, the concealed portions with a fan of thirteen striæ, and the basal margins with twelve distinct

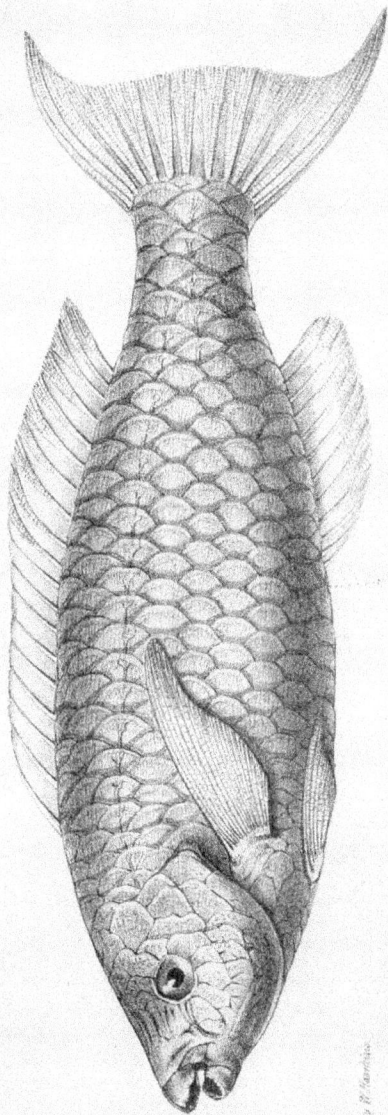

Scarus Chloris Rüpp.

Drawn from Nature by P. Gardner

crenatures. Lateral line interrupted; its first portion at the depth of two and a half rows of scales beneath the dorsal, and stopping beneath the commencement of the soft part of that fin; recommencing three rows lower down, exactly in the middle of the depth, whence it runs straight to the caudal.

Dorsal commencing above the opercle; the spinous portion of nearly uniform height, and scarcely more than one-fifth of the depth; the soft portion much higher, and terminating in a sharp point behind. Anal answering to the posterior half of the dorsal, terminating opposite to it, and similarly pointed; with six spines, which, as well as the dorsal spines, are furnished with very conspicuous filamentous tags. Caudal nearly even. Pectorals rounded, but not very broad, their length rather more than three-fourths that of the head; the rays rather slender. Ventrals pointed; the first soft ray elongated, and reaching to the anal when laid back; the last ray attached at its base by a membrane to the abdomen. Rows of small scales between the rays of the caudal at the base of the fin; and a few small ones along the base of the dorsal and anal, more particularly on the soft portions.

Colour.—" Above, greenish black; the sides paler; slightly iridescent."—D.—In spirits it appears of a nearly uniform brown all over, fins included.

Habitat, Maldonado, Rio Plata.

Mr. Darwin obtained this species at Maldonado, in a lake of fresh water, said sometimes to be a little brackish. It appears to belong to the genus *Chromis* of Cuvier, placed by him amongst the *Labridæ*, but having evidently very strong affinities to some of the *Sciænidæ*. It differs essentially from the *C. Brasiliensis* of Quoy and Gaimard,* in having six anal spines, and being destitute of all markings and spots. I am not aware that it is described by any author.

1. Scarus chlorodon. *Jen.*

Plate XXI.

S. æruginoso cyaneus, capite et pinnis flavo-vittatis: maxillis exteriùs lævibus, marginibus crenatis; caninis ad angulos oris nullis: fronte gradatim proclivi: lineâ laterali tubis parum ramosis: squamis ubique striuto-granulatis: pectoralibus acuminatis, radiis superioribus arcuatis: caudali radiis externis cæteris longioribus, acuminatis.

D. 9/10; A. 3/9; C. 13. &c.; P. 15 vel 16; V. 1/5.

Long. unc. 16.

Form.—Of an oval form; the greatest depth one-third of the length, caudal excluded: dorsal and ventral lines equally convex. Head a little less than the depth of the body, not gibbous in front, but with the profile falling regularly and gradually from the commencement of the dorsal. Snout rather pointed. Jaws equal, their outer surface smooth, but crenated on their cutting edges. No spinous canines at the corners of the mouth. Eyes rather small, their diameter not one-fifth the length of the head, situate above the middle of the cheek, but equidistant from the posterior lobe of the opercle and the extremity of the snout. Snout in front of the eyes, and

* *Freycinet Voyage, (Zoologie)* p. 286.

the lips, naked; but the cheeks and opercular pieces covered with large scales, which form two rows on the cheeks. Opercle terminating behind in a rounded angle. Scales on the body very large; eight in the depth, and twenty-one or twenty-two in the length: the entire exposed portion of each scale scabrous with granulations, which are partially disposed in lines towards the free edges. No scales on the vertical fins. The lateral line occupies the second row of scales from the top, till it reaches a little beyond the end of the dorsal, where it becomes interrupted, recommencing in the fourth row, which at this point is the third: tubal pores in some places ramified, but the ramifications not very distinct.

The dorsal commences above the posterior lobe of the opercle, and is of nearly uniform height throughout. The length of the rays in the soft portion, which is slightly higher than the spinous, is not quite one-third of the depth. The whole length of this fin is half the entire length. The anal answers to the last half of the dorsal, and terminates in the same line ; the three spines are slender, and the first very short. Caudal with the central portion slightly convex, but the three outer rays above and below prolonged into a point one-third the length of the whole fin ; the lower point a little longer than the upper. Pectorals about one-fifth of the entire length, pointed, with the upper rays arcuate. Ventrals immediately beneath them, one-third shorter.

COLOUR.—" Fine verditer blue, with some yellow stripes about the head and fins."—D.—The dried skin is nearly of a uniform brown, but the snout and cheeks are much varied with green : the jaws also are green. A bright green patch in front of the eye, immediately beneath which is a pale frænum, probably yellow in the recent state. Dorsal and anal green: the former shews some trace of a lighter narrow band running longitudinally below the upper edge of the fin ; the latter exhibits a very distinct fascia running along the middle. Caudal pale green, with the upper and lower edges of a much deeper tint. Ventrals in like manner edged with green. Pectorals wholly dusky.

Habitat, Keeling Island, Indian Ocean.

In so extensive a genus as the present, and one in which so much general similarity prevails amongst the species, the task of determining whether any particular one has been described before is extremely difficult. I can only say that the species which I have here ventured to characterize as new has been carefully compared with the descriptions of all those noticed in the "Histoire des Poissons," and though there are several to which it is nearly allied, there is none to which it can be referred with certainty. It seems to approach nearest the *S. variegatus*, but that species is said to have the caudal square, by which I presume is meant that the upper and under rays are not prolonged into a point, as is the case in so many species of this genus, and in the one here described.

This species was taken by Mr. Darwin at the Keeling Islands.

2. SCARUS GLOBICEPS. *Cuv. et Val.*

S. globiceps, *Cuv. et Val.* Hist. des Poiss. tom. xiv. p. 179.

FORM.—Oblong-oval, very much compressed throughout: the dorsal and ventral lines nearly of equal curvature. Greatest depth contained about three times and one-third in the entire length: thickness twice and three-fifths in the depth. Head one-fourth of the entire length, rather elevated at the nape, the forehead convex, whence the profile descends nearly in the arc of a circle, giving the snout a blunt and rounded appearance. The height of the head, taken in a vertical line through the eyes, equals nearly but not quite its own length. Mouth small, the gape not reaching half-way to the eye. Jaws very slightly crenated on their cutting edges, the true teeth appearing on the outer surface like minute scales. At the posterior angle of each jaw, and on each side, are two sharp canines projecting horizontally from the corners of the mouth, eight in all. Eyes rather small, their diameter contained six-and-a-half times in the length of the head, situate a little above the middle of the cheek, and a trifle nearer the extremity of the snout than the posterior margin of the opercle. The nostrils consist of two minute orifices a little in advance of the eye, and a little distant from each other, the posterior one largest and kidney-shaped, the anterior round and nearly closed by its membranous border. A cluster of minute pores above and behind the eyes, and a few others scattered about the snout.

Scales on the body very large, increasing in size at the base of the caudal, where there are three very large ones covering the rays of that fin for half their length or more: twenty-three in a longitudinal line, and nine in the depth. Each scale of a roundish form anteriorly, the basal portion with a projecting lobe in the middle of the hinder margin, and with thirty-one striæ in the fan; the exposed portion finely striated and granulated, with a broad membranaceous border: those on the caudal nearly three times as long as broad, but the ordinary ones with the length and breadth nearly equal. Lateral line interrupted; the upper portion running nearly straight at about one-fourth of the depth, till opposite the end of the dorsal, where it inclines downwards: tubal pores very distinctly ramified.

Dorsal very low, its height, in the middle of its length, being scarcely more than one-eighth of the depth: the soft rays slightly higher than the spinous, and increasing in length backwards. Anal answering to the last half of the dorsal, and terminating in the same line: three spines at its commencement not stouter than the soft rays, the first very small. The last soft ray in both dorsal and anal double. Caudal with the points about one-fourth of the rest of its length; when spread, the interval is rectilineal, but when the rays are closed the whole appears crescent-shaped. Pectorals a little shorter than the head, of a somewhat triangular form, the rays gradually decreasing in length from the uppermost to the lowermost. Ventrals pointed, about two-thirds the length of the pectorals, and immediately beneath them. A large oblong lanceolate scale between the ventrals, nearly half their length: also an oblong scale in the axilla of each, equalling the last of the soft rays.

D. 9/10 ; A. 3/9; C. 13, &c.; P. 13; V. 1/5.

Length 11 inches.

Colour.—Not noticed in the recent state. In spirits, it appears bluish grey on the back and sides with small round whitish spots, the margin of each scale being defined by a purplish line; paler on the belly: a white transverse line in front of the eyes passing from one to the other; anterior part of the snout, mouth, cheeks, and lower part of the head, yellowish white. Dorsal and anal pale, the former with three narrow longitudinal purplish lines, the latter with one. A portion of the under surface of the pectorals, extending from the third to the fifth ray, and forming a longitudinal fascia, purple ; the rest of those fins, as well as the caudal and ventrals, pale or nearly colourless.

Habitat, Tahiti.

This species was taken by Mr. Darwin at Tahiti. It so nearly answers to the description of the *S. globiceps* of Valenciennes, brought by MM. Garnot and Lesson from the same locality, that I cannot suppose it to be distinct. This specimen, however, appears to have more spinous teeth at the corners of the mouth.

3. Scarus lepidus. *Jen.*

S. fuscus, capite et pinnis purpureo-cæruleo tinctis: fronte parum elevato, æque ac rostro continue et gradatim proclivi ; hoc apice obtuso: maxillis exterius lævibus, marginibus vix crenatis ; canino ad angulum oris in maxillá inferiore unico, in superiore nulla: lineá laterali distincté ramosâ : pectoralibus subtriangulis: cauduli subæquali, radiis externis mediis vix longioribus.

D. 9/10 ; A. 3/9 ; C. 13, &c. ; P. 13 ; V. 1/5.

Long. unc. 8. lin. 7.

Form.—General form not very dissimilar to that of the last species, but the crown and nape less elevated, whence the profile falls in a more gradual slope: snout, nevertheless, blunt at the extremity. Depth of the body very nearly one-third of the entire length. Head about one-fourth of the same. The height of the head is about four-fifths of its own length. Jaws smooth externally, the true teeth appearing like minute scales on their surface, the cutting edges scarcely at all crenated : only one laterally projecting canine at each corner of the lower jaw, none in the upper. Diameter of the eye one-sixth of the head.

Dorsal not quite so low as in the last species ; its height in the middle of its length about one-seventh of the depth. Caudal nearly even, the upper and lower rays being scarcely longer than the others. Pectorals and ventrals similar, but the scale between the latter shorter and more rounded. Scales on the body large, the free portions finely striated and granulated, with a broad membranaceous border: three large ones at the base of the caudal, as in the last species. Lateral line distinctly branched, the ramifications irregular and varying on each scale ; in some instances only one long stem extending nearly to the margin of the scale, with one or more lateral twigs ; in others, two, three, or even four distinct stems, either simple or ramified.

Colour.—(*In spirits.*) Of a nearly uniform dark brown, with some faint traces of purplish blue about the head and fins, which possibly may have pervaded some parts of the body also in the recent state.

Habitat, Tahiti.

This species was taken with the last, and notwithstanding it presents two or three obvious differences in respect of form, as well as of colour, it is just possible it may be the same in a younger state. I think it not improbable that the points of the caudal may elongate with age, the forehead become more gibbous, and the spinous teeth more numerous. If it be distinct it would seem to be undescribed ; though the colours not having been noticed in the recent state renders it difficult to speak with certainty on this point. For the same reason, the specific character may perhaps hereafter be found to require alteration.

4. SCARUS ———?

Mr. Darwin's collection contains another species of *Scarus* from the Keeling Islands, which may probably be distinct from all those hitherto noticed, but which being in rather a bad state of preservation, I shall content myself with describing as well as I can, without affixing any name to it, lest in the end it prove not new. Many of the species enumerated in the "Histoire des Poissons" having only their colours noticed, it requires that these should have been observed more in detail than what Mr. Darwin's notes furnish in this instance, in order to decide whether it be identical or not with any of those spoken of in that work.

FORM.—A tolerably regular oval, somewhat attenuated at each extremity: dorsal and ventral lines of equal curvature. Nape not at all elevated, and the profile on the whole falling very regularly and gradually from thence to the end of the snout, though there is a slight eminence on the forehead. Depth one-fourth of the entire length. Jaws smooth externally, but with the true teeth very distinct upon their surface, and much more so upon their cutting edges than in either of the last two species. One horizontally projecting canine at each corner of the upper jaw, but none in the lower. The terminating lobe of the opercle is slightly emarginated behind, the membrane projecting immediately above the notch in the form of a short salient point. Lateral line interrupted, the upper portion nearly straight, and not inclining downwards at its posterior extremity: the tubes very slightly ramified, and many of them quite simple. The scales on the body are very finely granulated and striated: there are no large ones at the base of the caudal. Dorsal and anal low: height of the former contained four and a half times in the depth of the body, and exactly equalling the distance from the upper edge of the back to the lateral line. Pectorals somewhat triangular, the uppermost ray of all a little arcuate. Scale between the ventrals one-third the length of those fins. Caudal slightly crescent-shaped, when the rays are closed: when spread, all the middle rays appear even, the uppermost and lowermost projecting very slightly beyond them.

D. 9/10 ; A. 3/9 ; C. 13, &c. ; P. 14 ; V. 1 5.

Length 6 inches.

COLOUR.—"Body dull reddish and greenish, the colours being blended and mottled: fins banded lengthwise with vermillion-red: head with waving bright green lines."—D.—No trace of bright colours remains in its present state, and the only indication of markings is a narrow crescent-shaped band across the middle of the caudal.

MALACOPTERYGII.

FAMILY. SILURIDÆ. .

1. PIMELODUS GRACILIS. *Val.*

Pimelodus gracilis, *Val.* in D'Orb. Voy. dans l'Amer. Mérid. Atl. Ichth. Pl. 2. fig. 5.

———————— *Cuv. et Val.* Hist. des Poiss. tom. xv. p. 134.

FORM.—Of a slender elongated form, the body compressed behind the dorsal. Greatest depth contained about seven and a half times in the entire length: thickness at the commencement of the dorsal a little less than the depth. Head, measured to the gill-opening, rather more than one-sixth of the entire length : its breadth two-thirds of its own length. Helmet smooth, and not very conspicuous, though with its whole surface finely wrinkled : its breadth behind the eyes rather more than one-third of its length, measuring this last from the end of the snout to the further extremity of the interparietal process. The solution of continuity extends back nearly to the base of the process just mentioned, which last is narrow and lanceolate, three times as long as broad at its base, but not reaching to the buckler, or triangular plate in front of the dorsal, by one-third of its own length. The buckler itself is not very large, but sufficiently obvious.

Profile sloping gradually downwards in nearly a straight line from the beginning of the dorsal to the end of the snout : this last depressed and rounded horizontally in the form of a semicircle. Mouth wide, but very little cleft, the commissure not reaching half way to the eye. Upper jaw projecting a very little beyond the lower. In each a band of very fine velutine teeth ; but none on the vomer or palatines. Tongue smooth, and fastened down all round. Six barbules ; the maxillary pair very long, reaching to the commencement of the anal fin ; of the submandibular pairs, the exterior reach one-third beyond the insertion of the pectorals ; the interior are only half the length of the exterior. Eyes round, of moderate size, their diameter four and a half times in the length of the head, situate in about the middle of the length: distance from one to the other one diameter and a quarter. Lateral line nearly straight throughout its course, dividing the body longitudinally into two nearly equal portions.

Pectorals not quite equalling the length of the head, and a little less than one-sixth of the entire length : the spine very little shorter than the soft rays, very strong, with sharp teeth on its inner edge, but the outer edge only granulated, or with a few slight serratures towards the extremity. The humeral bone seen above the pectoral projects backwards in the form of a spinous lamina, but does not appear through the skin ; it equals half the length of the pectoral itself. The dorsal commences at one-fourth of the entire length, and is of a somewhat rectangular form, the soft rays not decreasing much backwards : its length equals four-fifths of the depth of the body, and two-thirds of its own height. The spine is not so strong as that of the pectoral, and with only a few small serratures on the outer edge near the tip. The space between the dorsal and the adipose a little exceeds the length of the former. The adipose

itself is twice the length of that fin; very low at first, but gradually rising, until, before its
termination, it becomes equal to between one-half and one-third of the depth. The vent is in
the middle of the entire length, caudal excluded. Anal short, and just beneath the middle of
the adipose, there being about one-fourth of this last fin in advance of it as well as behind it :
the first four rays simple, but apparently all articulated, the first two or three very minute and
not easily observed. Caudal forked for two-thirds of its length: the upper lobe a little longer
than the lower, and contained five and a half times in the entire length. Ventrals immediately beneath the last ray of the dorsal ; a little shorter than the pectorals, and not reaching
to the anal by half their own length.

D. 1/6 ; A. 14 or 15 ; C. 17, &c. ; P. 1/9 ; V. 6.

Length 5 inches 2 lines.

COLOUR—(*In spirits.*) Brownish, inclining to silvery in some places : a dusky fascia formed of
dots along the lateral line. Dorsal rather dusky at the base, and with the upper portion also
dusky between the rays: a dusky spot on the anterior part of the adipose.

Habitat, Rio de Janeiro.

This species was taken by Mr. Darwin in a running brook at Rio de Janiero.
It approaches on the whole so nearly the *P. gracilis* of D'Orbigny, that I can
hardly suppose it to be distinct. Yet there are some slight differences observable in this specimen. It has more anal rays ; the adipose appears shorter ; and
the upper lobe of the caudal is not so prolonged, though possibly it may be worn
down. Also D'Orbigny's figure appears to want the dusky stains on the dorsal
and adipose fins. If it be not that species it must be new, as there is none other
described by Cuvier and Valenciennes with which it will assimilate better.

2. PIMELODUS EXSUDANS. *Jen.*

*P. corpore parum elongato, altitudine quintam partem longitudinis æquante: galeâ
lævi, inconspicuâ, processu interparietali haud clypeum parvum prædorsalem attingente : poris paucis buccalibus amplis, serie obliquâ dispositis : maxillis æqualibus :
cirris sex ; maxillaribus haud anulem attingentibus : lineâ laterali primum deflexâ,
deinde rectâ : pinnis dorsali et anali brevibus ; adiposâ dorsali haud duplo longiore :
caudali profundè bifurcâ, lobis æqualibus : spinâ pectorali margine interno fortiter
dentato.*

D. 1/7 ; A. 13 vel 14 ; C. 17, &c. ; P. 1/8 ; V. 6.

LONG. unc. 3. lin. 6.

FORM.—In some respects resembling the last species, but the body much less elongated, the depth
and thickness remaining the same. The depth is about one-fifth of the entire length; the
head rather more than one-fifth. The helmet is scarcely so much wrinkled, and the interparietal process not so long, reaching only half-way to the buckler, which last is smaller and
less obvious. The solution of continuity of the bones of the cranium appears to extend back

in the form of a narrow fissure nearly to the base of the interparietal process, but is not very obvious, except between the eyes, where it opens into a sort of elongated ellipse. There are eight or nine pores on the top of the head, rather behind the eyes, so arranged as to form nearly a complete circle. There is also a very conspicuous row of three or four large oval pores on the cheek, at the anterior part of the opercle, descending obliquely forwards: other smaller ones may be seen scattered about different parts of the head. The jaws are equal: the teeth as in the last species, excepting that a roughness can be distinctly felt upon the vomer beneath the skin, though there are no teeth on that part which appear through it. The maxillary pair of barbules reach to a point midway between the insertion of the ventrals and the commencement of the anal: of the submandibular pairs, the exterior do not reach beyond the insertion of the pectorals; the interior are two-thirds the length of these. Eyes elliptical, the longitudinal diameter about one-fifth the length of the head; situate nearer the end of the snout than the posterior part of the opercle; the interval between them nearly two diameters. The lateral line slopes downward till opposite the fourth soft ray of the dorsal, then passes off straight along the middle to the caudal.

The pectorals are contained about five-and-a-half times in the entire length. The spine is similar to that of the last species; but the humeral bone is shorter, and scarcely one-third of the pectoral itself. The dorsal is similar; and the space between it and the adipose is the same; but the adipose itself, though of the same form, is not so long, from the body being less elongated; it is not more than half as long again as the dorsal. Anal similar, but the adipose not advancing so much beyond it. Caudal and ventrals similar; but the former with the lobes equal.

COLOUR.—(*In spirits.*) On the whole similar to, but darker than that of the last species. There is some appearance of a fascia along the lateral line. The upper part of the dorsal, and also of the anal, is dusky, but there is no spot on the adipose. Pectorals and ventrals dusky.

A *second specimen* differs from the above in no respect, except in being smaller, measuring two inches and a quarter in length, and in having one ray more in the anal.

Habitat, Rio de Janeiro ?

The number annexed to this species has been lost, but there is reason for believing that Mr. Darwin took it with the last at Rio de Janeiro. I cannot identify it with any of those described by Cuvier and Valenciennes in the "Histoire des Poissons."—It affords another instance of the indefiniteness of the character derived from the presence or absence of vomerine teeth; since a band of such teeth, which are considered by Valenciennes as absent in this genus, can be distinctly felt and made obvious by removing the skin of the palate, though they do not appear externally. This is not the case with the species last described, to which nevertheless, the present one approaches so closely in general character, that they never could be widely separated. Except for the greater elongation of the body in the *P. gracilis*, I should have been almost inclined to consider them as sexes of the same species.

Callichthys paleatus. *Jen.*

C. capite compresso lævi: ore parvo, cirris maxillaribus ad utrumque angulum duobus, haud ultrà oculos pertingentibus; labro inferiore reflexo, et in duos cirros breves membranaceos producto: spinâ pectorali compressâ, forti, margine interno leviter dentato, externo lævi, quintam partem totius longitudinis æquanti: caudali bifurcâ, lobis æqualibus acuminatis.

D. 1/7—1 ; A. 6 ; C. 14, &c. ; P. 1/7 ; V. 6.

Long. unc. 1. lin. 10.

Form.—General form resembling that of the *C. punctatus*. Depth, at the commencement of the dorsal, one-third of the length, excluding caudal: thickness at the pectorals three-fourths of the depth. Head slightly compressed, its height very little less than its length, this last, measured to the gills, being rather less than one-fourth of the entire length. Profile falling from the dorsal in one regular slope, and nearly rectilineal till it arrives before the eyes, where it curves downwards, making the extremity of the snout obtuse. Mouth small, the upper jaw a little projecting; two maxillary cirri at each angle; these nearly equal; the lower one a little the longest, reaching to beneath the middle of the eye: also two short cirri, only half the length of the maxillary ones, pendent from the reflexed lower lip, a little separate from each other, one on each side the middle. Teeth so minute as hardly to be distinguished; a row can just be felt on each jaw, and on the vomer. Head smooth. The number of dorsal laminæ twenty-one; that of the ventral twenty.

Pectorals a little exceeding the length of the head; the spine nearly as long as the fin itself, but not exceeding one-fifth of the entire length; very strong, compressed, and sharp-pointed, with a few fine teeth or serratures on the inner margin, but smooth on the outer. Height of the dorsal more than half the depth, and a little more than its own length, this last equalling the space between it and the adipose: the spine strong, and similar to that of the pectoral. Anal directly underneath the adipose, and hardly occupying more space. Ventrals shorter than the pectorals, attached beneath the last third of those fins, or under the second soft ray in the dorsal, and scarcely reaching more than half-way to the anal: the first ray, as well as that of the anal, somewhat hispid. Caudal forked for half its length, which about equals that of the head: the lobes equal and pointed.

Colour.—(*In spirits.*) General colour yellowish-brown, with dusky spots and mottlings: breast and edge of the abdomen whitish. Pectorals, ventrals, and anal, almost wholly dusky; dorsal and caudal spotted.

This species, in its general character, approaches so extremely near the *C. punctatus* of Valenciennes, that perhaps I am in error in considering it distinct. But it is remarkable for possessing, in addition to the four usual maxillary cirri, two labial, which are neither represented in D'Orbigny's figure,* nor noticed in the description given in the "Histoire des Poissons," and which therefore I infer are not present in that species, as they could hardly have been overlooked, or been deemed unimportant to be noticed. The maxillary cirri also, which in the *C. punctatus,*

* *Voy. dans L'Amér. Mérid. Atl. Ichth. pl. 5. fig. 1.*

Q

reach, according to Valenciennes, to the gill-opening, if not beyond it, here only attain to beneath the middle of the eye ; and this character is invariable in five specimens which Mr. Darwin has brought home. Judging from the description, there would seem to be one or two further differences : the profile appears to be more rectilineal, the pectoral spine shorter, and smoother on its external margin. The colours are on the whole similar, but the pectorals and ventrals darker : the latter, which are said to be yellow in the *C. punctatus*, are here quite dusky in every one of the specimens.

The exact locality in South America in which Mr. Darwin obtained this species is uncertain, as the specimens have lost their attached labels.

Family.—CYPRINIDÆ.

1. Pœcilia unimaculata. *Val.*

Pœcilia unimaculata, *Val.* in Humb. Zool. et Anat. Comp. vol. ii. p. 158. pl. 51. fig. 2.

Form.—Body oval, slightly elongated, thick anteriorly, compressed behind. The dorsal and ventral lines meeting at the mouth at an acute angle ; but the head, when viewed from above, broad, and very much flattened between the eyes, and the snout obtuse. Greatest depth about one-fourth of the entire length: thickness two-thirds of the depth. Length of the head nearly equalling, or a little less than, the depth of the body. Mouth small : jaws very protractile ; each with a single row of very fine, close-set, pointed teeth ; the lower one a trifle the longest. Eyes large, their diameter three and a half times in the length of the head, high in the cheeks, reaching to the line of the profile. Nostrils consisting of one small orifice a little above and rather in advance of the eyes.

Scales large, investing the head and all the pieces of the gill-cover, though very thin and transparent on the opercle and not very obvious there. On the body there are about eight in the depth, and twenty-seven or twenty-eight in a longitudinal row from the gill-opening to the caudal. One taken from the middle of the side found to be of a semi-elliptic form, the exposed portion marked with numerous very fine curved concentric lines, the basal with sixteen or seventeen deeper-cut nearly parallel striæ gradually lengthening from the sides towards the middle, but not converging to a fan. Lateral line very faintly marked out by a dotted line, scarcely obvious in some places.

Dorsal small, commencing exactly at the middle point of the entire length, measuring this last quite to the extremity of the caudal. Anal similar and opposite; in strictness, however, terminating a very little in advance. The last ray in both these fins double: the first two in the anal short. Caudal rounded. Pectorals and ventrals small and narrow, the former three-fourths the length of the head ; the latter not above half the same. The pectorals, when laid back, reach to the insertion of the ventrals, but the ventrals hardly reach to the commencement of the anal.

B. 5 ; D. 7 ; A. 9 ; C. about 24, including short ones ; P. 14 or 15 ; V. 6.

Length 2 inches.

COLOUR.—Greenish-brown, with a conspicuous black spot on the middle of each side, a little in
advance of the commencement of the dorsal. Dorsal a little dotted and mottled with dusky,
especially towards the tips of the rays. The other fins plain.

Habitat, Rio de Janeiro.

This species, which was discovered by Humboldt, was observed by Mr.
Darwin in great numbers in fresh-water ditches at Rio de Janeiro: others were
taken in equal plenty in a salt lagoon. The bellies of the females are very tur-
gid when big with young, which are said to be excluded alive, and yellowish.—
Valenciennes, in his description, speaks of the opercle as being smooth, or without scales, though he says the preopercle is covered with scales; and he would
lead one to suppose that they are absent on this part in the whole genus, as it
enters into his generic character; I find them, however, present, though very thin
and transparent, both in this species and the next.

The general resemblance which *Pœcilia* bears to *Mugil*, as regards the form
of the head and mouth, is very striking, and calls up irresistibly the idea of
some relation of analogy between these two genera.

2. PŒCILIA DECEM-MACULATA. *Jen.*

PLATE XXII. Fig. 1.

*P. corpore sub-elongato, viridescenti-fusco; lateribus maculis nigris circiter decem
serie longitudinali dispositis; pinnis immaculatis: dentibus subincisivis: caudali
subtruncatâ.*

D. 8; A. 10; C. 22, brevibus inclusis; P. 9; V. 5.

LONG. unc. 1. lin. 4.

FORM.—More elongated than the last species; the snout not so acute when viewed laterally.
Depth not more than one-fifth of the entire length, the length of the head being equal to it.
Mouth and jaws similar; the teeth also in one row in each jaw, and forming a compact series,
but more incisor-like than pointed, with oblique cutting edges. Scales of a different form
and sculpture; more oblong than semi-elliptical, broader than long; the deep striæ behind
more numerous, amounting to twenty or more, and all drawn nearly of the same length. They
cover all the pieces of the opercle as in the *P. unimaculata.*

Dorsal and anal exactly opposite, commencing at a point a little anterior to the middle of
the entire length, reckoning this to the extremity of the caudal. Caudal rather more approaching to square than rounded; the number of rays fewer than in the last species.
Pectorals narrower, having also fewer rays. Ventrals very small, scarcely more than half
the length of the pectorals. When laid back, the pectorals reach to beyond the insertion of
the ventrals: the ventrals do not attain to the anal.

COLOUR.—Greenish-brown, with about ten conspicuous somewhat oval-shaped dusky spots,
arranged in a longitudinal line along the middle of each side. All the fins plain.

Habitat, Maldonado.

This, which is evidently a new species of *Pœcilia*, was taken by Mr. Darwin at Maldonado, in a lake that had been suddenly drained. There are three specimens in the collection, none of them exceeding the length above given. Mr. Darwin, however, states in his notes, that he believes them to be full grown, having taken them so repeatedly, in brooks, of the same size. The number of spots varies from nine to twelve, and is sometimes different on the two sides of the same specimen.

Independently of the spots, which at once characterize this species, it is readily distinguished from the last by its teeth, which are more cutting than pointed, and in this respect rather departing from the character of the genus as established by Valenciennes.

1. LEBIAS LINEATA. *Jen.*

PLATE XXII. Fig. 2.

L. corpore subelongato, subrompresso, viridescenti-fusco ; lateribus lineis circiter septem longitudinalibus nigris, e maculis parvis subconfluentibus formatis : dentibus uniserialis : caudali rotundatâ.

D. 9 ; A. 9 ; C. 26, brevibus inclusis ; P. 13 ; V. 6.

LONG. unc. 1. lin. 10.

FORM.—General form very similar to that of the *Pœcilia decem-maculata.* Slightly compressed ; the depth one-fifth of the length ; the length of the head about four-and-a-half times in the same. Head depressed : snout obtuse : mouth small ; the commissure horizontal. Upper jaw very protractile ; the lower one rather the longest, when the mouth is shut. Teeth forming a single closely-set series, somewhat compressed at bottom, the cutting edges tricuspid. Diameter of the eye nearly one-fourth the length of the head. Some large conspicuous pores on the lower jaw, passing upwards in a series along the margin of the preopercle, not very near together, about eight or nine in all.

Scales large, covering the head and all the pieces of the gill cover, as well as the body. About eight in the depth, and thirty in a longitudinal line from the gill to the caudal. One taken from the middle of the side of a semi-elliptic somewhat oblong form ; the free portion very finely striated, the basal with ten or twelve deeper-cut striæ, these last nearly parallel, and of equal lengths. Lateral line faintly marked out by a dotted line ; the first half in the third row of scales from the top, the last half in the fourth row.

Dorsal commencing at exactly the middle point of the entire length. Anal opposite and similar. Caudal rounded. Pectorals small, about two-thirds the length of the head. Ventrals smaller, barely one-half of the same. The pectorals, when laid back, reach to the insertion of the ventrals ; but the latter hardly attain to the anal.

COLOUR.—Greenish-brown, with six or seven longitudinal dark lines on the sides, the lines apparently made up of spots for the most part confluent, but here and there not so, interrupting the continuity of the lines. All the fins pale dusky, without any spots or markings.

Habitat, Maldonado.

This new species of *Lebias* was taken by Mr. Darwin in the same lake at

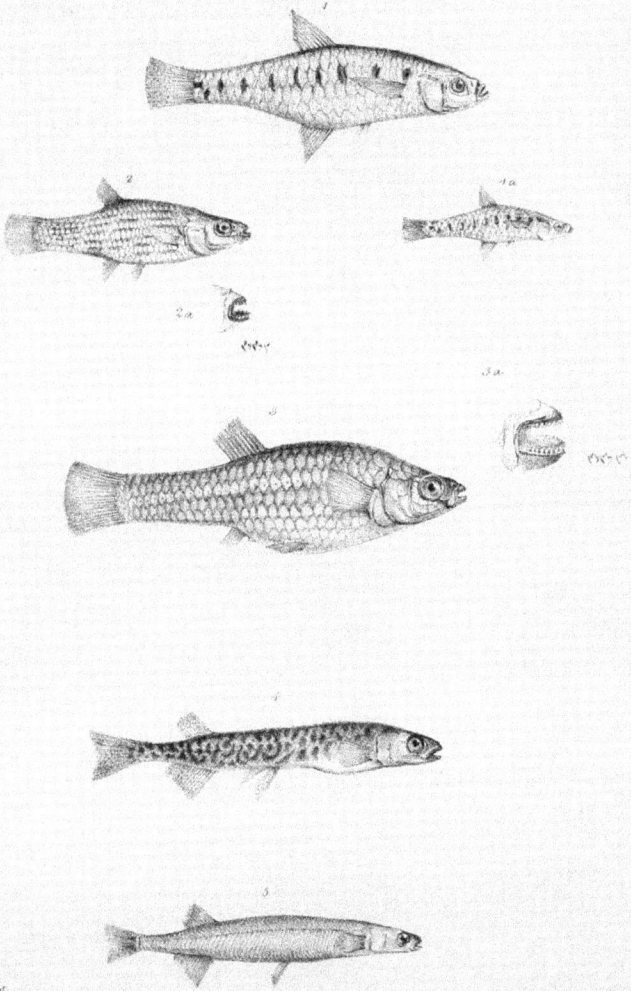

From Nature on Stone by W. Hawkins.

No. 1 Pœcilia decem-maculata. *Magnified View twice Natural Size*

1a. .... *Natural Size*

2. Lebias lineata .... *Nat. Size*

2a. .... *Magnified View of Teeth*

3 Lebias multidentata .... *Nat Size*

3a. .... *Magnified View of Teeth*

4. Mesites maculatus)
5. .... attenuatus) *Nat. Size*

Maldonado with the *Pœcilia decem-maculata*. There are several specimens in the collection, none of them exceeding the size above mentioned, and they have all the appearance of being full grown. Some have the lines of spots much more interrupted than others.

2. LEBIAS MULTIDENTATA. *Jen.*

PLATE XXII. Fig. 3.

L. corpore subelongato, subcompresso, viridescenti-fusco ; lateribus fusciis angustis paucis longitudinalibus albidis obscurioribus : dentibus seriebus plurimis dispositis, omnibus tricuspidatis : caudali rotundatá.

D. 9 ; A. 9 ; C. 26, brevibus inclusis ; P. 13 ; V. 6.

LONG. unc. 3 lin. 2.

FORM.—The general form and proportions of this species are extremely similar to those of the last; but it differs very remarkably in having behind the anterior row of tricuspid teeth, a band of minuter teeth above and below, all of which are also tricuspid, and similarly formed to those in front. Head one-fifth of the entire length ; flattened on the crown. Jaws nearly equal ; upper one very protractile. Scales large ; about thirty-two in a longitudinal line, and eight in the depth ; covering all the pieces of the opercle ; similar in form to those of the last species, but with the striæ on the free portion finer and more numerous, the deep-cut basal striæ also rather more numerous, amounting to about fourteen, and of unequal lengths, gradually increasing from the outermost to the middle ones. Lateral line similar ; also the same pores on the lower jaw. Fins and finray-formula similar : in both species the first and last rays of the dorsal and anal are simple, and shorter than the others. The anal perhaps terminates a little nearer the caudal than the dorsal does.

COLOUR.—(*In spirits.*) Greenish-brown, with very little appearance of markings in its present state. There is, however, some indication of an irregular scattered row of small black spots on each side, a little below the ridge of the back ; also of two or three pale longitudinal narrow bands along the middle of the sides, which were probably more conspicuous in the living fish. The belly is yellow, and very tumid; but these are evidently characters merely indicative of the female sex.

Habitat, Monte Video.

This is another new species of *Lebias* taken by Mr. Darwin in fresh-water at Monte Video, if indeed it strictly belong to the genus ; but the circumstance of the teeth being in several rows, and in fact forming a complete band, is at variance with the generic characters as given by Cuvier. The teeth however being exactly of the same form as in the other species, and the general characters on the whole similar, I have not thought it expedient to erect it into a new genus. There is but one specimen in the collection, which appears to be a large female big with young.

Genus—MESITES. *Jen.*

Corpus elongatum, gracile, antice subcylindricum, postice compressum, nudum, squamis nullis. Caput depressum. Rostrum breve, obtusum: os terminale, rictu modico. Maxillæ debiles; superior margine ex ossibus intermaxillaribus omnino formato, maxillaribus retroductis et a labio partim celatis. Dentes minuti, acuti, in maxillâ utrâque uniseriati; in linguâ et vomere biseriati; in ossibus palatinis et pharyngalibus nulli. Apertura branchialis amplissima, membranâ sex-radiatâ, subter gulam profundè emarginatâ, haud isthmo annexâ. Pinnæ dorsalis et analis valde retropositæ, oppositæ. Pinnæ pectorales et ventrales parvæ. Pinna caudalis leviter emarginata.

There can be no doubt, I imagine, as to this being an entirely new form, and a very interesting one, from the circumstance of its being at the extreme end of the family to which it belongs, and its very much departing from the usual characters of that family. I have referred it to the *Cyprinidæ*, taking that group in the enlarged view in which Cuvier accepts it ; though by those who divide it into subfamilies it would probably be associated with the *Cobitidæ*, or made to constitute a distinct one by itself. It agrees with the *Cyprinidæ* in general in the form of its mouth, in the upper jaw having its margin entirely formed by the intermaxillary, the maxillary being present, but placed behind and partly concealed in the thickness of the lip, and in the want of an adipose ; but it altogether departs from that family in the entire want of scales, of which there is not even a vestige in the dried skin, and in which respect it would seem to shew an affinity to the *Siluridæ*. Yet it has none of the other characters of the family just mentioned. On the other hand, in the backward position of the dorsal and anal fins, which are opposite to each other, it agrees with the *Esocidæ*. The pharyngeal bones are unarmed, but this deficiency is made up for by the strong curved teeth on the tongue, independently of the minuter ones in the jaws.

The intestine is extremely short and quite straight, measuring only fourteen lines in length from the pylorus to the anus, in a specimen two inches and a half long. The stomach is of an oval form, of considerable capacity, very membranaceous, with the cardiac and pyloric openings near together at the upper extremity, from the latter of which the intestine is immediately reflexed to pass off to the anus. In the specimen dissected, the stomach was much distended by a nearly perfect individual of the genus *Colymbetes*, which appeared to have been recently swallowed, and was scarcely at all altered. There are no cœcal appendages. The air-bladder is of an elongated oval form, and of considerable development.

Mr. Darwin's collection contains no less than three species of this new genus, differing but slightly from each other. Two are from the most southern parts of South America, the third from New Zealand.

1. MESITES MACULATUS. *Jen.*

PLATE XXII. Fig. 4.

M. viridescenti-fuscus; dorso et lateribus maculis crebris, hic et illic confluentibus, nigris ; ventre niveo ; pinnarum radiis nigro-punctatis.

B. 6; D. 10 ; A. 16; C. 16, &c. ; P. 12 ; V. 7.

LONG. unc. 2. lin. 8.

FORM.—Slender and very much elongated. Body anteriorly subcylindrical, compressed behind. Greatest depth not more than one-eighth of the entire length : thickness about three-fourths of the depth. Head rather depressed, about one-sixth of the entire length. Snout short and rounded ; mouth at the extremity; the gape moderate, not quite reaching to beneath the anterior angle of the eye. Lower jaw ascending a little to meet the upper, and, when the mouth is open, appearing rather the longest. Intermaxillary fixed, forming the entire margin of the upper jaw, the maxillary being behind it, and, though of nearly equal development, not very distinct: both bones slender. Teeth small, but sharp-pointed, rather widely apart, arranged in a single row along the edge of the intermaxillary, and in the lower jaw; the series above consists of about eighteen, that below of about twenty-one : also a double longitudinal row on the tongue, each row containing five or six teeth, the anterior ones curved, and larger than any of those in the jaws : a similar double row, but of minuter ones, down the middle of the vomer; none, however, on the palatines or pharyngeans. Eyes rather large, their diameter contained about three and a half times in the length of the head, distant scarcely one diameter from the end of the snout. The nostrils appear to consist of only a single aperture in front of the eye, in the neighbourhood of which, and also above the eye, are several large pores. The opercle and subopercle taken together approach to an oblong form, the posterior margin being straight and nearly vertical : the subopercle is not much developed, nor very distinct. Gill-opening very large, the membrane thick, with six rays, deeply notched beneath, and not fastened down. The whole skin is perfectly smooth and naked, invested with mucosity. No appearance of any lateral line, unless a fine dark streak be so called, passing along the middle of the sides, and dividing them into two equal parts.

The dorsal and anal are opposite to each other, and both placed very far back, almost at the extremity of the body. They commence in nearly the same vertical line, a very little anterior to the commencement of the last third of the entire length ; but the anal being longer than the dorsal, it extends nearer the caudal. The form of these fins is much as in the genus *Cobitis*. The dorsal has the first three rays simple, the rest branched : the anal also has the first three simple, the first very short. Caudal about one-eighth of the entire length, with a shallow notch, the principal rays branched. The vent is just before the anal. The ventrals arise from about the middle of the entire length, the distance from their insertion to the commencement of the anal being twice their own length. The pectorals are small, and rather narrow, equalling about two-thirds the length of the head or hardly so much : they are attached low down, but not quite so low as in the genus *Cobitis*.

COLOUR.—(*In spirits.*) Greenish-brown, with numerous conspicuous spots and small irregular transverse bars of black. Under a lens the spots appear to be made up of thickly crowded black specks upon a dark brown ground : the bars result from some of the spots being confluent. The belly appears to have been white. The rays of all the fins are dotted with dusky, but the membranes transparent and colourless.

The individual described above was taken by Mr. Darwin in a fresh-water brook, in Hardy Peninsula, Tierra del Fuego. His collection, however, contains four other specimens found in streamlets and creeks high up the river of Santa Cruz in Patagonia, where they are said to have been numerous. Though these last are slightly different, they are evidently referable to the same species : they also vary a little from each other. Their peculiarities are as under :

The largest measures 2 inches 8 lines in length, and has the following fin-ray formula :

D. 12; A. 16; C. 16, &c.; P. 14; V. 7.

The next in size is 2 inches 6 lines, with the fin-ray formula thus :

D. 11; A. 16; C. 16, &c.; P. 13; V. 7.

These specimens agree in being both slenderer than the one from Tierra del Fuego. The depth is eight and a half, if not nine times in the entire length : the head rather more than one-sixth of the same. The colours are similar, except that the spots are not quite so numerous, and of a more regular form, seldom running together to form bars.

It is to these specimens that Mr. Darwin's notes refer, respecting the colours of this species in the recent state. As follows : " Pale greenish brown, with small irregular transverse bars of black ; belly snow white."—D.

The third of the Patagonian specimens is 2 in. 2 lin. long. Fin-ray formula—

D. 10; A. 15; C. 16, &c.; P. 14; V. 7.

The fourth is of the same length.

D. 11; A. 15; C. 16, &c; P. 13; V. 7.

These last two specimens are exactly similar to each other in colours, but differ from the former two in being almost immaculate, having only a few spots on the upper part of the back. This brings them very close to the following species, from which they are scarcely to be distinguished, except by their smaller eyes. It should be observed further, that the fleshy part of the tail in these specimens has the upper and under edges fringed with the short accessory rays of the caudal, a character which is not so obvious in any of the others.

2. MESITES ALPINUS. *Jen.*

M. viridescenti-fuscus, dorso saturatiore ; hoc, et lateribus, et pinnis, nigro levissimè irroratis, immaculatis ; ventre niveo ; oculis majusculis.

D. 10 ; A. 16 ; C. 16, &c. ; P. 13 ; V. 7.

LONG. unc. 2. lin. 5.

FORM.—Very little difference in form between this and the last species. The eyes, however, are decidedly larger, measuring in diameter one-third the length of the head. The head itself also appears somewhat longer, being nearly one-fifth of the entire length. The anterior teeth on the tongue do not seem much larger than the others. Fins similar.

COLOUR.—(*In spirits*). Greenish brown, deepening in tint at the top of the back. Back, sides, and fins, immaculate, but thickly powdered with minute dark specks, scarcely visible except under a lens. These specks give the fins a more dusky appearance than they possess in the last species. The belly appears to have been white.

A *second specimen* does not differ from the above in any respect, except in being rather smaller, and scarcely more than two inches in length.

Habitat, Tierra del Fuego.

This species was taken by Mr. Darwin in alpine fresh-water lakes in Hardy Peninsula, Tierra del Fuego. I have no hesitation in considering it distinct from the last, as there are two specimens exactly similar, both shewing a larger eye, and an entire absence of all approach to spots ; while the whole surface of the back and sides is thickly irrorated with dark specks, a character which does not appear in the plain varieties of the *M. maculatus.*

3. MESITES ATTENUATUS. *Jen.*

PLATE XXII. fig. 5.

M. viridescenti-fuscus, ventre vix pallidiore ; dorso, et lateribus, pinnarumque radiis, sparsim nigro levissimè irroratis, immaculatis : corpore prægracili, anticè attenuato ; capite et oculis minoribus.

D. 11 ; A. 17 ; C. 16, &c. ; P. 12 ; V. 7.

LONG. unc. 2. lin. 6.

FORM.—Rather more slender than either of the last two species, the body more attenuated anteriorly ; the head also smaller, though scarcely shorter. Mouth and eyes both smaller ; the diameter of the latter not more than one-fourth of the length of the head. Teeth also rather more minute as well as more numerous. The fins are similar, except that the ventrals appear to stand rather more forward, being attached exactly in the middle of the entire length, excluding caudal : the distance from their insertion to the commencement of the anal is more than twice their own length. The outer rays of the caudal are worn, but there was probably a shallow notch when entire : the short accessory rays are very numerous, and form a very distinct fringe along the upper and lower edges of the fleshy part of the tail.

R

COLOUR.—(*In spirits.*)   Greenish-brown, much more uniform than in either of the last two
species, not deepening on the back, and scarcely becoming paler underneath.   The back,
sides, and rays of the fins, are finely irrorated with dusky specks, as in the *M. alpinus*, but not
to the same extent, the specks being more thinly scattered, and here and there scarcely
visible.   From the same cause the fins appear paler.

Habitat, Bay of Islands, New Zealand.

This, which is a very distinct species of this new genus, was taken by Mr.
Darwin in fresh-water in the Bay of Islands, New Zealand.   It is well characterized by its more attenuated head and smaller eye, than those of either of the
two others.

FAMILY.—ESOCIDÆ.

EXOCŒTUS EXSILIENS. *Bl. ?*

Exocœtus exsiliens, *Bl.* Ichth. pl. 397.

FORM.—Head about one-sixth of the entire length, and approaching to the form of a paralle-
lopiped ; very much flattened on the crown and between the eyes quite to the end of
the snout, broader above than beneath, so that the cheeks are beyond the vertical inclining
inwards at bottom.   Snout short : mouth not much cleft ; when shut, the jaws are equal, and
the commissure of the lips appears to extend to beneath the anterior margin of the eye, but
the maxillary, which retires completely beneath the suborbital, does not reach so far : when
the mouth is open, the maxillary becomes vertical, and the intermaxillary being scarcely at all
protractile, the lower jaw is a little the longest.   Teeth very minute : a row, scarcely visible,
along the forepart of the intermaxillary, but not extending to the sides of the jaw : none that
can even be felt in the lower jaw, or in any other part of the mouth.   Tongue rounded, and
free at the tip.   A loose veil of skin hangs down in front of the palate, from immediately
behind the teeth in the upper jaw.   Eyes round, and very large ; the upper part of the orbit
reaching to the line of the profile, and forming a slightly salient ridge : their diameter very
nearly one-third the length of the head ; between them and the end of the snout is two-
thirds of a diameter ; the distance from one to the other across the crown is one diameter and
a quarter.   The nostrils consist of one large round orifice a little in advance of the eyes.
The membrane of the opercle forms a slightly salient angle backwards, near the upper part of
the gill-opening.   Scales large, of a somewhat irregular form, approaching to oblong, nearly
twice as broad as long, the posterior margin with three or four incisions near the middle, and a
few rather indistinct nearly parallel striæ on the surface of the basal portion ; in others these
striæ converge to form a small but very regular fan ; and the scales appear to vary a good deal
on different parts of the body.

The pectorals reach exactly to the base of the lateral caudal rays ; the first two rays are
simple, and all the others branched ; first ray of all not half the length of the fin.   Dorsal so
situate as to leave a space between it and the end of the fleshy part of the tail about equal to
its own length ; the first ray simple, the others branched ; the last prolonged beyond those
which immediately precede it so as to form rather a point backwards.   Anal similar to the

B. Waterhouse Hawkins del.

No. 1 Tetragonopterus Abramis.
2 _____ rutilus
3 _____ scabripinnis.
4 _____ interruptus.

} All Nat. Size.

1a. 2a. 3a. 4a. Magnified View of Teeth.

dorsal, and answering to it exactly. The ventrals are attached at a point, in this specimen, about half an inch posterior to the middle of the entire length, reckoning this to the end of the fleshy part of the tail; but are not much out of the middle, if the length be reckoned to the end of the upper lobe of the caudal: their length is contained not quite three and a half times in the entire length, excluding caudal; and they reach exactly to the end of the dorsal and anal : first ray very much branched, and only one-third the length of the fin; second ray appearing like two or even three rays at its upper extremity, from the circumstance of the several branches of it being of unequal length; all the other rays branched likewise. Vent a very little in advance of the anal. Upper lobe of the caudal one-third shorter than the lower; this last being exactly the same length as the ventrals.

D. 11 ; A. 12 ; C. 16, &c. ; P. 18 ; V. 6.

Length 12 inches 8 lines.

COLOUR.—The colours were not noticed in the recent state; and the specimen is in such bad condition, and so much altered by the spirit, that they are now no longer distinguishable.

The flying-fish above described was taken by Mr. Darwin in the Pacific Ocean, off the coast of Peru, in Lat. 18° S. It appears to be the *E. exsiliens* of Bloch, but as I am not aware that the species of *Exocœtus* have been ever rigourously worked out, and closely compared from different parts of the globe, I have thought it expedient to annex a description, by which it may be identified, if it prove hereafter distinct.*

FAMILY.—SALMONIDÆ.

1. TETRAGONOPTERUS ABRAMIS. *Jen.*

PLATE XXIII. fig. 1.

T. corpore subrhomboideo, compressissimo, altitudine fere dimidium longitudinis, pinnâ caudali exclusâ, æquante : osse maxillari angusto, retrorsum arcuato : pinnâ dorsali triangulari, suprà ventrales accuratè exorienti ; anali lævi, longâ, altitudine retrò cito decrescente ; utráque plicis membranaceis, radiis longitudinaliter adhærentibus, instructâ : squamis in lineâ laterali 46, in lineâ inter pinnas ventrales et dorsalem transversâ 17.

B. 4 ; D. 1/10 ; A. 2/30 ; C. 19, &c. ; P. 13 ; V. 8.

LONG. unc. 4. lin. 7.

FORM.—Of a subrhomboidal form, the nape and back being much elevated, whence the profile falls very obliquely and in nearly a straight line. Greatest depth nearly half the entire length, excluding caudal. Body very much compressed, the thickness being nearly three and a half times in the depth. Head approaching to a laterally flattened cone, with the length and height nearly equal. Snout very short; mouth but little cleft; when open, the lower jaw projecting

* Swainson is of opinion that " more than double the number of species of *Exocœtus* really exist above those that have been described."—*Nat. Hist. of Fishes*, vol. i. p. 299.

considerably. Maxillary narrow, and of nearly equal breadth throughout, curving backwards.
Teeth with their cutting edges dentated, the middle point much the most developed, with one
or two smaller ones on each side: two rows of such teeth on the intermaxillary, and one in
the lower jaw, this last row with scarcely more than eight or ten teeth in it. No teeth on the
maxillary, vomer, palatines, or tongue. Eyes round, rather large, their diameter three and a
half times in the length of the head, distant not so much as one diameter from the end of the
snout. Nostrils with two orifices, the posterior one a narrow curved slit, the anterior one a
round hole. The suborbital forms a somewhat triangular naked disk beneath the eyes, with
radiating veins. Posterior margin of the opercle very little curved: subopercle narrow, and
small, forming but a small portion of the gill-flap.

About seventeen scales in the depth, and forty-six in the lateral line, which last bends
downwards rather below the middle, and is continued quite to the caudal. A scale taken from
the middle of the side below the lateral line is somewhat rounded anteriorly, the basal margin
being straight ; the surface marked with very fine numerous concentric striæ, and with two
coloured deeper striæ on the free portion diverging from the centre in a V-like form : some
scales have three or four of these coloured striæ, drawn more or less regularly.

The dorsal commences in about the middle of the entire length, excluding the caudal
and narrow part of the tail; of a triangular form, its greatest height equalling the depth to the
lateral line. Pectorals narrow, shorter than the head, attached below the bottom of the gill-
opening, and reaching rather beyond the insertion of the ventrals, which last are in a vertical
line with the commencement of the dorsal and shorter than the pectorals. Anal long, commencing a very little beyond the tips of the reclined ventrals; the anterior portion rather more
than half the height of the dorsal, but the posterior much lower, sloping rapidly off; two spines,
the first very minute, the second about one-third the length of the first soft ray; the last soft ray
double. The anal terminates nearly in a line with the adipose, which is small. Caudal forked
for half its length : the lobes equal. Many of the rays in the vertical fins, more especially the
dorsal and anal, are accompanied through nearly half their length from the bottom by membranous folds of skin. There are also some small scales along the base of the anal, but
none apparent on the dorsal. In the axillæ of the ventrals is an elongated scale, not half
their length.

COLOUR.—"Back bluish silvery, with a silver band on the side: a bluish black spot behind the gills.
Fins pale orange ; tail with a black central band."—D.—There is now not much trace of the
silver band, or the black band on the tail. The humeral spot is, however, still very distinct.

Habitat, the Rio Parana, South America.

This species was taken by Mr. Darwin in October in the Rio Parana, as high
up as Rozario. I cannot ascertain that it is described, though there is much resemblance between it and the species figured in Seba.* It differs from the *T.
chalceus* of Spix, in its much smaller scales, not to mention other points of dissi-
milarity.

* Thesaurus, vol. iii. pl. 34. f. 3.

2. TETRAGONOPTERUS RUTILUS. *Jen.*

PLATE XXIII. fig. 2.

T. corpore ovali, compressissimo, altitudine tertiam partem longitudinis, hâc ad basin furcæ caudalis mensâ, æquante: osse maxillari angusto, retrorsum arcuato: pinnâ dorsali subtriangulari, paululum pone ventrales exorienti; anali lævi, longâ, altitudine retrò cito decrescente; utrâque plicis membranaceis, radiis longitudinaliter adhærentibus, instructâ: squamis in lineâ laterali 40, in lineâ inter pinnas ventrales et dorsalem transversâ 14.

D. 1/9 ; A. 2/27 ; C. 19, &c. ; P. 14 ; V. 8.

LONG. unc. 4. lin. 3.

FORM.—More oval than rhomboidal: the back and nape not so much elevated as in the last species; the profile falling less obliquely. Depth one-third of the entire length, measured to the base of the caudal fork: head one-fifth of the same. Not above fourteen scales in the depth, and forty in the lateral line, which occupies the eighth row from the top. The scales on the whole similar, but with the basal margin not so straight and regular, and somewhat projecting in the middle in the form of a blunt salient angle. The dorsal commences a trifle nearer the caudal, and at the middle of the entire length, the caudal alone excluded. The pectorals reach just to the insertion of the ventrals, which last are a trifle in advance of the dorsal. The second anal spine is longer, and nearly half the length of the soft rays which follow.

COLOUR.—"Back iridescent greenish brown: a silver band on the side. Fins dirty orange: tail with a central black band; above and below the band bright red and orange."—D.—The colours appear very similar to those of the last species. The humeral spot, however, is less obvious, while, on the other hand, the silver band on the side can still be distinguished.

Habitat, Rio Parana, South America.

Taken with the last species, to which it is very closely allied. Mr. Darwin observes in his notes, that both are among the commonest of the river fry in the Rio Parana.

Both this and the *T. Abramis* are distinguished by having narrow longitudinal folds of skin attached to the basal half of the rays of the dorsal and anal fins, a character which does not shew itself in any of the three species next to be described.

3. TETRAGONOPTERUS SCABRIPINNIS. *Jen.*

PLATE XXIII. fig. 3.

T. corpore ovali, subelongato, valde compresso; altitudine tertiam partem longitudinis, caudali exclusâ, æquante: osse maxillari paulo dilatato, recto: pinnâ dorsali suboblongâ, altâ, pone ventrales exorienti; anali scabrâ, altitudine retrò gradatim decrescente; radiis plicis membranaceis nullis: squamis in lineâ laterali circiter 38, in lineâ inter pinnas ventrales et dorsalem transversâ 12.

D. 1/9 ; A. 3/22 ; C. 19, &c. ; P. 13 ; V. 8.

LONG. unc. 3. lin. 7.

Form.—Still more oval and elongated than the last species, the profile falling in a gentle curve from the dorsal to the end of the snout. Depth exactly one-third of the length, excluding caudal: head one-fifth of the entire length, caudal included. Maxillary rather dilated towards the bottom, and quite straight, not curving backwards as in both the last species. Teeth rather larger, and more conspicuous. Nostrils larger. Only twelve scales in the depth, the lateral line occupying the seventh row from the top: thirty-seven or thirty-eight in the lateral line. Scales of a rather different form; the basal margin more sinuous, the free margin not so regularly curved, the coloured striæ hardly obvious. Dorsal more oblong than triangular, higher in relation to the depth, of which it equals two-thirds, commencing at a point anterior to the middle of the length, excluding caudal. The pectorals reach a little beyond the ventrals, which are attached a little in advance of the dorsal. The anal has all the rays longer, and more nearly equal, the posterior part of the fin not being so much sloped off: there are three spines at the commencement, the first two minute, the third not quite half the length of the soft rays: this fin is furthermore distinguished from that of the two former species by the rays being set with asperities, which communicate a scabrous harsh feel to the touch, when the finger is passed along them from the base upwards.

Colour.—Not noticed in the recent state. In spirits it appears more silvery than either of the two last species: the back and upper part of the sides being brownish. A humeral dusky spot, and the remains of what was probably a bright silver band along the middle of the side from the gill to the caudal. At the base of the caudal is a dusky spot, which is prolonged in a line along the central rays to the commencement of the fork. The other fins faintly edged with dusky, but otherwise pale.

Habitat, Rio de Janeiro.

The more oval and elongated form, straight maxillary, and scabrous anal fin, at once distinguish this species from either of the two last. It was taken by Mr. Darwin in fresh water, at Rio de Janeiro, in June.

4. Tetragonopterus tæniatus. *Jen.*

T. corpore ovali, valde compresso, altitudine tertiam partem longitudinis, hâc ad basin furcæ caudalis mensâ, æquante: osse maxillari margine posteriore recto: pinnâ dorsali suprà ventrales accuratè exorienti; anali lævi, altitudine retrò cito decrescente; radiis plicis membranaceis nullis: squamis in lineâ laterali 40, in lineâ inter pinnas ventrales et dorsalem transversâ 14.

D. 1/10 ; A. 3,22, &c.

Long. unc. 2. lin. 2.

Form.—Depth and general form similar to those of the *T. rutilus;* also the same number of rows of scales, the lateral line occupying the eighth from the top. Anal smooth, and similar to that of the *T. rutilus* in form, but in the number of the spines and soft rays agreeing with the *T. scabripinnis.* The maxillary straight, but hardly so much dilated as in the last-named species, being of nearly equal breadth throughout. The teeth are more numerous than in either, amounting in the lower jaw to fourteen or more. The ventrals are in an exact line with the commencement of the dorsal.

COLOUR.—Much as in the *T. scabripinnis.* The same silver band, only more brilliant; also the same humeral spot, and the spot at the base of the caudal extending along the middle rays.

Habitat, Rio de Janeiro.

The above description is that of two small specimens, similar to each other, obtained by Mr. Darwin in a running brook at Socego, in the province of Rio de Janeiro. They are probably not full grown ; but so evidently distinct from either of the last two species, the characters of which are in some measure combined in them, that I have not hesitated to give them a separate place. The silver band is more or less obvious in all the species of this genus brought home by Mr. Darwin, but it is much brighter in this than in any of the others.

5. TETRAGONOPTERUS INTERRUPTUS. *Jen.*

PLATE XXIII. fig. 4.

T. corpore ovali, valde compresso : altitudine tertiam partem longitudinis totius æquante : ore parvo ; osse maxillari brevissimo, dilatato, margine posteriore recto : dentibus minutis, multicuspidatis : dorsali subtriangulari, pone ventrales exorienti ; anali (in uno sexu ?) scabrá, altitudine retrò subito decrescente ; his pinnis plicis membranaceis nullis : squamis in lineá longitudinali 35, in lineá inter pinnas ventrales et dorsalem transversá 10 : lineá laterali interruptá, haud finem pinnæ pectoralis attingente.

D. 1/10 ; A. 2/18 ; C. 18, &c ; P. 11 ; V. 7

LONG. unc. 2. lin. 8.

FORM.—Oval, and not very dissimilar in general form to the *T. rutilus*, but rather more elevated above the shoulder. Depth exactly one-third of the entire length: head one-fourth of the same, caudal excluded. Profile not exactly straight, but very slightly hollowed out at the crown, then as slightly protuberant above the eyes, whence it falls more rapidly in front, giving the snout a short and blunt appearance. Mouth much smaller than in any of the preceding species, owing to the extreme shortness of the maxillary, which is broad, with the anterior margin curving outwards, but the posterior one straight. Teeth very small ; the points on the cutting edges numerous (five or six on each tooth) and nearly equal ; apparently only one row on the intermaxillary, and the same in the lower jaw ; none on the maxillary. Eyes and nostrils as in the other species, but the anterior orifice of the latter larger. Subopercle much larger, forming a greater portion of the gill-flap. Scales relatively larger; only ten in the depth, and thirty-five in the length. They have no deep striæ on the basal portion, and consequently no fan ; but they are very regularly marked with the usual finer striæ concentrically arranged, except on the free portion where they become indistinct.

The lateral line occupies the sixth row of scales from the top, but is very soon interrupted, coming to an end before it has reached the length of the pectoral, and not carried over more than eight or nine scales in the length. Dorsal subtriangular, commencing exactly at the middle of the length, caudal excluded. Anal shorter than in the other species, and not reaching so near the caudal ; two spines, but the first extremely minute. Caudal forked for half its length, the lobes equal. Ventrals more forward than in the other species, decidedly in advance of the dorsal, and attached at one-third of the entire length; their axillary scale very small. Pectorals narrow, reaching beyond them. No long folds of skin accompanying the rays of the vertical fins.

COLOUR.—Not noticed in the recent state, in which, judging from its appearance in spirit, it was probably silvery, or perhaps golden, with somewhat of an olivaceous hue becoming deeper on the back. There are evident remains of a bright longitudinal lateral band : also of a black spot at the base of the caudal. The dorsal and the anterior portion of the anal incline to dusky : the pectorals and ventrals are slightly dusky at their extremities; there is also a large irregular dusky stain on the back and shoulders anterior to the dorsal fin.

A *second specimen* differs from the above in being a trifle smaller, and not quite so deep in the body. The anal is also decidedly scabrous, as in the *T. scabripinnis*, and has one ray less in it. The colours are similar, excepting that the fins are rather less dusky, and the large dusky stain on the back and shoulders is wanting.

Habitat, Maldonado.

This species is one of several that were taken by Mr. Darwin at Maldonado, in the lake that had been suddenly drained, before alluded to. It is immediately distinguished from all the others in this genus above described, by its small mouth and abbreviated lateral line. The circumstance of the anal fin being scabrous in only one of the specimens leads to the suspicion that this may be a sexual character, perhaps common to this and several species; and, judging from its somewhat less depth, I conceive the specimen so distinguished in this instance to be a male.

HYDROCYON HEPSETUS. *Cuv.*

Hydrocyon hepsetus, *Cuv.* Reg. An. (ed. 2) tom. ii. p. 312.
———— falcatus, *Freycinet*, (Voyage) Zoologie, p. 221, pl. 48. fig. 2.

FORM.—Back rising slightly from the nape, whence the profile in front falls obliquely in nearly a straight line to the mouth. Depth contained about three and a half times in the length, caudal excluded. Both head and body much compressed, the greatest thickness being only two-fifths of the depth. The length of the head equals the depth of the body. Snout appears rather pointed when the jaws are shut ; when open, the lower jaw is a little the longest. Gape considerable. Maxillary long, commencing before the eyes, and reaching to a vertical from the posterior part of the orbit ; inclining downwards, lapping obliquely in part over the lower jaw, gradually widening towards the posterior extremity, which is rather obliquely rounded. Intermaxillary with two sharp canines in front, then on each side four or five very small hooked teeth, then another large canine, though not so large as those in front; behind this commences the maxillary, which is armed all along its margin with a regular row of small equal hooked teeth, resembling sharp serratures ; a similar row on each palatine, but none on the vomer or tongue; this last pointed, and free at the tip. Lower jaw with two strong canines in front, larger than those in the upper, and fitting into two holes above, when the mouth is shut ; on each side of these are three only half their size, but increasing backwards, placed at rather wide intervals; then follow a row of close, minute, sharp teeth, similar to those on the edge of the maxillary. Eyes rather large, their diameter not quite one-fourth the length of the head, distant one diameter and a quarter from the end of the snout. Suborbital large, consisting of three pieces. Preopercle rectangular. A row of pores, not very distinct, along the under part of the lower jaw, thence continued along the limb of the preopercle. Opercle and sub-opercle taken together with the posterior margin forming a slight but regular curve, with scarcely any salient angle.

Head naked; scales on the body of moderate size, arranged in somewhat oblique rows, especially below the lateral line; one from the middle of the side below the lateral line of an irregularly rounded form, the posterior margin rather sinuous, the disk with numerous fine concentric striæ, but no deeper-cut striæ on the basal portion. About sixteen scales in a vertical row, and fifty-seven or fifty-eight in the lateral line: this last bending downwards in a curve which falls below the middle of the depth. Scales on the lateral line not larger than the others.

The dorsal answers to the space between the ventrals and anal; its height equals the depth to the lateral line. Anal long, commencing exactly under the last ray of the dorsal; the first part of this fin as high as the dorsal, but the rays, beyond the fifth, gradually decreasing; three spines, the first two very minute; the last soft ray double. Caudal in this specimen injured. Adipose and last ray of the anal in the same vertical line. Pectorals two-thirds the length of the head, attached very low down beneath the terminating portion of the gill-flap, narrow and slightly falcate, reaching to the ventrals, which last are one-fourth shorter. A long narrow scale in the axilla of each ventral one-third the length of the fin itself.

B. 4; D. 11; A. 3/26; C. 22, &c.; P. 12; V. 8.

Length 4 inches 3 lines.

Colour.—"Bluish silvery."—D.—Some appearance of a dusky spot at the base of the caudal prolonged in a line along the middle rays, but scarcely any trace of a humeral one. The dorsal and anal incline a little to dusky.

Taken at Maldonado, in a fresh-water lake, in June. I have scarcely any doubt of its being the *H. falcatus* of the Zoology of Freycinet's voyage, the figure of which it exactly resembles, excepting that the humeral dark spot, if it ever existed, and which is not mentioned in Mr. Darwin's notes, is now almost entirely effaced. It is probable, however, that there are two or three species nearly allied, for which reason I have been the more particular in my description. The *H. Hepsetus* of D'Orbigny* appears to differ from the *H. falcatus* of Freycinet (with which last Cuvier associates his name of *Hepsetus*,) in having the lateral line curving upwards rather than downwards, and the caudal fascia as well as the humeral spot more marked. The *Salmo falcatus* of Bloch is probably distinct from both.

---

* *Voy. dans L'Amer. Mérid.* Atl. Ichth. pl. 9. fig. 2.

s

Genus.—APLOCHITON. *Jen.*

Corpus elongatum, compressum, subfusiforme, undique nudum alepidotum.   Caput parvum.   Rostrum breviusculum, subacutum.   Os terminale, rictu modico.   Maxilla superior margine ex ossibus intermaxillaribus omnino formato, maxillaribus, hæc subæquantibus, retroductis.   Dentes minuti, acuti, in maxillâ utrâque uniseriati, in linguâ et vomere longitudinaliter biseriati, in ossibus palatinis nulli.   Apertura branchialis amplissima, membranâ triradiatâ, subtus profundè emarginatâ.   Pinnæ, dorsalis paululum pone ventrales, analis paululum pone dorsalem reclinatam, exorientes.   Ventrales appendicibus axillaribus nullis.   Pinna caudalis bifurca.

I have already noticed a remarkable new form among the *Cyprinidæ* brought home by Mr. Darwin, and differing from all the known genera in that family by the entire absence of scales. The one now to be described is not less remarkable among the *Salmonidæ*, and, what is particularly interesting, would seem to occupy an exactly analogous place in this family, departing from it in the same important character of having the skin perfectly naked and free from scales. There are, however, many other points of similarity between *Mesites* and the genus which I have here termed *Aplochiton.** In both there is the same form of mouth, the margin of the upper jaw being entirely formed by the intermaxillary, behind which is the maxillary of nearly equal development. The teeth in the jaws are similar, both in regard to form and arrangement; there is also the same double longitudinal row on the tongue, and along the vomer. The pieces of the opercle are similar, and the gill opening equally large in both genera, though the branchial membrane has twice the number of rays in *Mesites* that it has in *Aplochiton.* Furthermore, the fins are on the whole very similar, with the exception of the dorsal not being so far back in *Aplochiton,* and there being also an adipose in this genus. It is also deserving of notice that both these new forms, so resembling each other in many of their characters, come from the same quarter of the globe, being found either in the most southern parts of S. America, or in the neighbouring islands.

From the circumstance of the naked skin, *Aplochiton* might by some be referred to the *Siluridæ,* but what was said of the genus *Mesites* may be repeated here, that it has none of the other external characters of that family. The maxillary, instead of being reduced to a mere vestige, or lengthened into a barbule, is as much developed as in any of the *Cyprinidæ,* and of the usual form ; the subopercle also is very distinct ;† while there is no strong spine at the commencement of either the dorsal or pectoral fins. At the same time it must be mentioned that *Aplochiton*

---

* Ab απλοος simplex, et χιτων tunica.

† Valenciennes says, in his preface to the fifteenth volume of the " Histoire des Poissons," that none of the *Siluridæ* have the subopercle ; and that the absence of this bone serves to distinguish them from *Cobitis.*

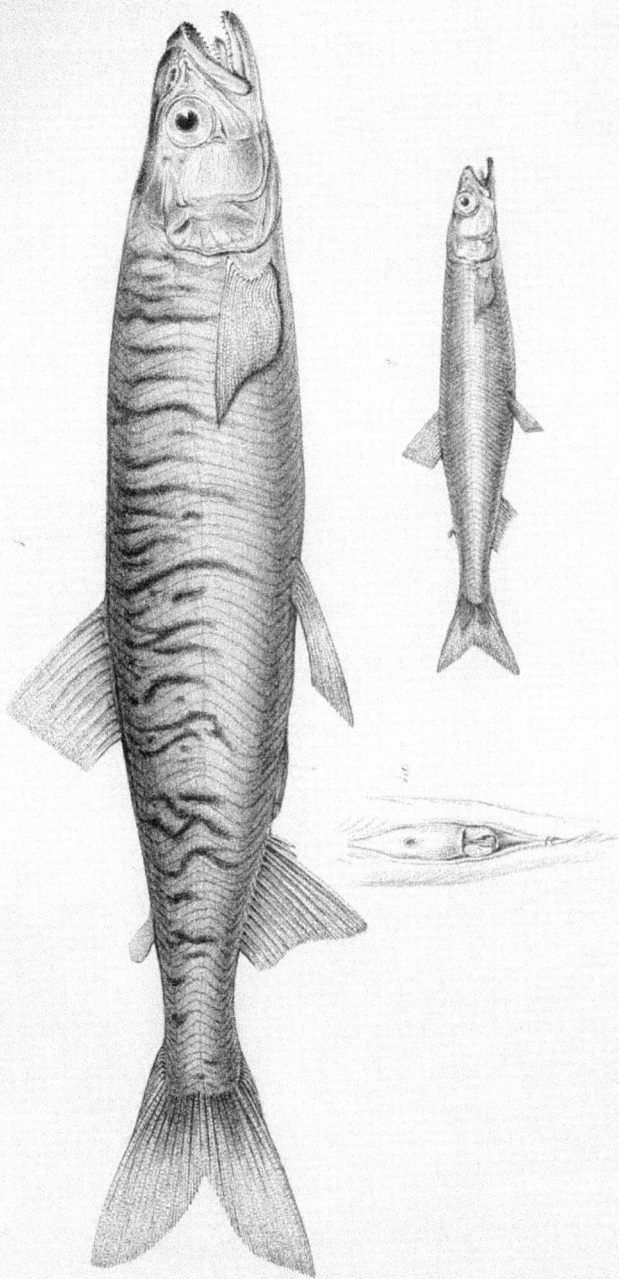

1. *Aplodactus Zebra.* Nat. Size.
1a. Magnified view of anal and genitourinary orifice
2. *Aplodactus lineatus.* Nat. Size.

agrees with the *Siluridæ* in having no cœcal appendages, though the air-bladder is similar to that of the ordinary *Salmonidæ*.   There are some peculiarities connected with the anal and sexual orifices which I shall notice presently, along with other points in the internal structure, in the species to be described first.

Mr. Darwin has brought home two species of this new genus, agreeing precisely in form, but very distinct in respect to size and colouring.

1. APLOCHITON ZEBRA.   *Jen.*

PLATE XXIV. FIG. 1.

A. obscurè plumbeus ; fasciis nigris transversis: maxillis æqualibus.

B. 3 ; D. 11 ; A. 2/14 ; C. 16, &c. ; P. 18 ; V. 7.

LONG. unc. 9. lin. 6.

FORM.—General form somewhat resembling that of the *Mackarel*, elongated, and approaching to fusiform.   Greatest depth about the middle, equalling one-sixth of the entire length.   Line of the back and profile nearly straight, the latter falling very little.   Greatest thickness rather more than half the depth.   Head small, contained about five and a half times in the entire length.   Snout short, but rather acute.   Mouth with a moderate gape reaching to beneath the anterior angle of the eye: when shut, both jaws equal, the lower one ascending at an angle of 45° to meet the upper; when open, the lower one a little the longest.   Margin of the upper jaw formed by the intermaxillary, the maxillary appearing behind it.   A single row of small but sharp teeth extending along the entire margins of both jaws: a double row of similar teeth, but stronger and more curved, down the middle of the tongue; also a double row along the middle of the vomer.   Eyes moderate; their diameter four and a half times in the length of the head; distant about one diameter and a quarter from the end of the snout, and with an interval between them of about one and a half.   Nostrils with two orifices, both roundish, one before the other, with a little interval between, the posterior one rather the largest.   Two distinct pores on each side of the crown, one behind the other with an interval between, above and rather behind the eyes.   Gill-opening very large, the membrane with only three flattened rays, deeply notched beneath, the notch reaching to beneath the middle of the eyes.   All the pieces of the opercle present, but the interopercle only just appearing behind the angle of the preopercle, and the subopercle forming but a narrow lanceolate lamina beneath the true opercle, which last constitutes the greater portion of the gill-flap, and is of an oblong form, the posterior margin being cut straight and vertical.

The whole skin perfectly naked everywhere, without the least vestige of scales.   No lateral line, except a faint streak, passing along the middle of the sides, be so called.   Dorsal commencing at the middle of the length, this last being measured to the base of the caudal fork; of the same form as in the ordinary species of the genus *Salmo ;* its greatest height a little exceeding its length, which last is rather more than half the depth of the body; first ray simple, the rest branched.   Adipose small, and just half way between the end of the dorsal and the base of the caudal.   Anal of a somewhat triangular form, the margin sloping very much off backwards, commencing a little beyond the tip of the reclined dorsal, and terminating opposite the adipose, or perhaps a trifle beyond it; two small spines at its commencement; the first two soft rays simple, the others branched. Caudal forked for half its length, the lobes equal; the whole fin contained about six and a half times in the entire length; the accessory rays very numerous, and partially fringing the upper and lower edges of the tail. Pectorals attached behind the gill-opening, rather below the middle, about two-thirds the length of the head, of a somewhat triangular form, the rays gradually shortening from the first, which is the only one unbranched. Ventrals attached a very little in advance of the dorsal, rounded, or almost cut square at the ends, the rays being all nearly equal. They are scarcely shorter than the pectorals: the space between their insertion and the commencement of the anal is nearly double their own length. There is no long scale or appendage of any kind in their axillæ.

Colour.—" Dull leaden colour."—D.—In spirits it appears brown. The sides are banded with some irregular transverse zebra-like marks, not noticed by Mr. Darwin, reaching from the back down two-thirds or three-fourths of the depth, some terminating sooner than others. All the fins brownish.

Habitat, Falkland Islands.

Mr. Darwin obtained three specimens of this remarkable fish all precisely similar, from a fresh-water lake in the Falkland Islands, in March. The lake was not far from the sea, and connected with it by a brook. He adds in his notes that the species is common there; that it is good eating, and grows to be about half as large again as the individuals procured.

One of these specimens was dissected by Mr. Yarrell and myself, and presented the following internal characters, which are of importance to be noted. The coats of the stomach were thick and muscular; the œsophageal portion with prominent longitudinal plicæ. Its contents, so far as they could be ascertained, consisted of the remains of caddis-worms. The intestine was large, without any cœcal appendages, but with one spiral convolution at the end of the first third of its length from the pyloric orifice : the entire length of the canal was four inches. The air-bladder was large, undivided, and of the same general form as in many of the *Salmonidæ*. There were two elongated flattened lobes of roe nearly ready for exclusion. The anal and sexual orifices were separated, but enclosed in a tubular sheath, common to both, directed backwards; the sheath itself lying in a groove in the abdomen, and five-eighths of an inch in length : the opening to the cavity of the abdomen and sexual organs was at the extreme end of this sheath, and partly closed by two lateral valves ; the opening to the intestine, three-eighths of an inch short of the extremity.

2. Aplochiton tæniatus. *Jen.*

Plate XXIV. Fig. 2.

A. olivaceus, punctis fuscis minutissimis irroratus; lateribus vittâ longitudinali argenteâ : maxillâ inferiore longiore.

3

B. 3; D. 12; A. 2/13; C. 16, &c.; P. 18; V. 7.

Long. unc. 3. lin. 10.

Form.—A much smaller species than the last, but the general form similar. Rather more elongated, the depth being contained seven and a half times in the length. Head one-fifth of the length measured to the base of the caudal fork. Snout a little longer, and more pointed. Lower jaw at all times a little the longest. Teeth similar, and similarly disposed. Nostrils similar; but no pores on the crown, or only one on each side, and that not very distinct. All the fins similar.

Colour.—Not noticed in the recent state. In spirits it appears of a uniform greenish or olivaceous brown, the back and sides very minutely dotted with darker brown. There is a pale silver band along the middle of the side, not bounded, however, by any definable line, but shading off insensibly into the brown above and below. The irides are still bright, and appear to have been golden.

Habitat, Goree Sound, Tierra del Fuego.

There are three specimens of this second species in the collection, all of the same size, and not differing in any respect from each other, except that one of them has thirteen rays in the dorsal fin, and fourteen soft rays in the anal. Mr. Darwin's notes state that they were taken at Goree Sound, Tierra del Fuego, in the mouth of a fresh-water stream, where the water was quite fresh; and that when put into salt water they immediately died.

The silver band at once distinguishes this elegant species from the last, independently of its smaller size. The specimens appear full grown.

There is the same peculiarity with respect to the anal and sexual orifices in this species, as in the one previously described.

Family.—CLUPEIDÆ.

1. Clupea Fuegensis. *Jen.*

Mr. Darwin's collection contains a single individual of a species of *Clupea* from Tierra del Fuego so extremely similar to the young of the common Herring, that it might almost be mistaken for it. As it is small, and in not very good preservation, I shall merely point out some of its leading characters.

Depth of the body the same as in a young *Herring* of the same size. Belly carinated, but with the serratures not more developed than in that species. Teeth the same, and very minute. The maxillary does not slope inwards quite so much at its upper extremity, before uniting with the intermaxillary; but the mouth and its several parts are in all other respects similar. The same may be said of the pieces of the opercle, excepting that there is a more sensible notch near the upper part of the posterior margin of the gill-flap, much as in the common *Sprat*. It differs also from the Herring in having the ventrals exactly in a line with the commencement of the dorsal, this last being placed a little further back than in that species.

D. 18; A. 19; C. 19, &c.; P. 18; V. 8.

Length 3 inches.

"Caught at night, off Cape Ines, Tierra del Fuego, two miles from the shore, in thirteen fathoms."—D.—The specimen is probably not full-sized,

2. Clupea arcuata. *Jen.*

The present species is from Bahia Blanca. The specimens, of which there are two, are also in bad condition like the last, and probably not full-sized.

Form.—Body deep, with the ventral line swelling rather more outwards than the dorsal. Greatest depth a little exceeding one-fourth of the entire length. Very much compressed: abdomen carinated, and very sharply serrated, the serratures sharper than in the common sprat. A few minute teeth at the extremity of the lower jaw, and also on each side of the intermaxillary, near its junction with the maxillary; the lower half of this last finely serrated. Pieces of the gill-cover much as in the Sprat; the subopercle rounded at bottom, the opercle with a shallow notch near the upper angle.

The dorsal commences exactly in the middle of the entire length, excluding caudal. The ventrals are as nearly as possible directly beneath its first ray: these fins are very small, and shorter than in the sprat.

D. 18; A. 23; C. 19, &c.; P. 16; V. 7.

Length 4 inches 2 lines.

Colours.—"Back blue; belly silvery."—D.

The second specimen is similar, only smaller. Both were taken in the month of September.

3. Clupea sagax. *Jen.*

A third species of *Clupea*, in Mr. Darwin's collection, likewise in bad condition, much resembles in general form the common Pilchard.

Depth the same as in the Pilchard, but the head larger and longer than in that species, being one-fourth* of the entire length. Abdomen smoother; no appearance of any serratures in front of the ventrals. Lower jaw but little advanced beyond the upper. No perceptible teeth, more than a few very minute serratures near the lower extremity of the maxillary. Diameter of the eye about one-fifth the length of the head. The form and veinings of the pieces of the opercle very much as in the Pilchard, but the interopercle more developed. The posterior margin of the opercle and subopercle taken together is almost quite straight, without any emargination anywhere, and not far out of a vertical; the subopercle is cut nearly square at bottom. The preopercle is much veined: there are also some veins on the upper part of

* In the pilchard it is one-fifth.

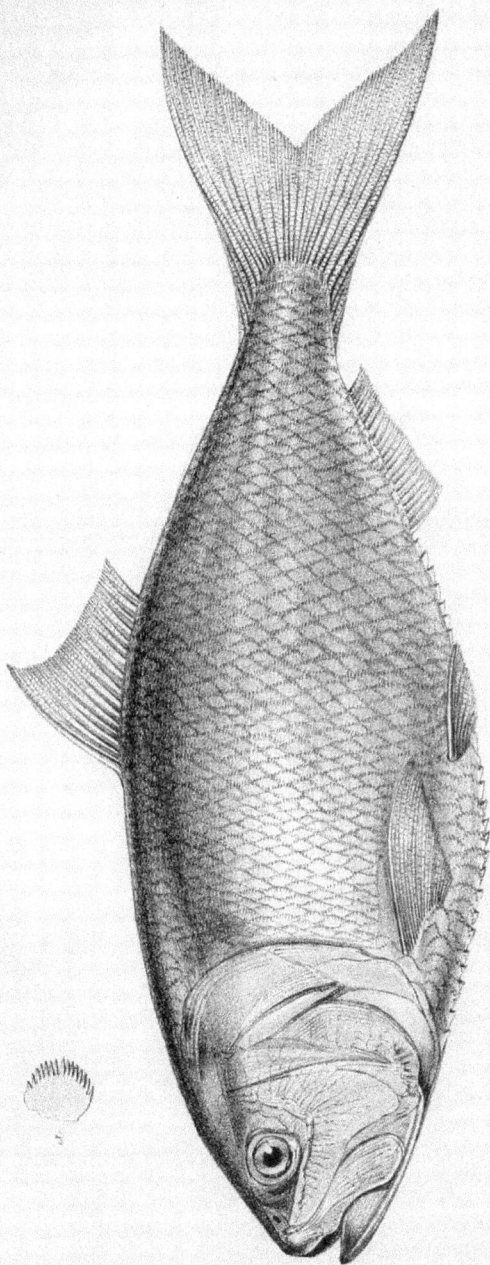

Alosa pectinata 4. Nat size.
a. Magnified. Scale from nape.

the opercle, and lower down on this last piece some very deep striæ, running parallel to its junction with the preopercle, as in the Pilchard, but still more strongly marked.

The commencement of the dorsal is very little anterior to a middle point between the end of the snout and the base of the middle caudal rays. The ventrals are beneath the posterior half of the dorsal as in the Pilchard. There are the same two elongated scales on each side of the caudal as in that species. The scales on the body, however, are much smaller than in the Pilchard, with their free portions striated, the basal portions marked with some irregular curved lines running in a transverse direction towards the median line of the scale.

D. 11 ; A. 18 or 19 ; C. 19, &c. ; P. 18 ; V. 8.

Length 10 inches 6 lines.

Habitat, Lima, San Lorenzo Island.

ALOSA PECTINATA. *Jen.*

PLATE. XXV.

A. corpore ovali, altitudine prope tertiam partem longitudinis æquante: ventre carinato, serraturis, præsertim inter pinnas ventrales et analem, acutis: maxillis subæqualibus, edentulis: preoperculo venoso ; operculo striato: squamis pectinatis : pinnis ventralibus paulo ante dorsalem exorientibus.

D. 16 ; A. 21 ; C. 19, &c. ; P. 17 ; V. 7.

LONG. unc. 12.

FORM.—Of an oval compressed form, the depth very considerable, equalling very nearly one-third of the entire length. Head contained about three times and three quarters in the same. Abdomen sharply carinated, with strong serratures, especially between the ventrals and anal. Jaws nearly equal, perhaps the lower one a little the longest; intermaxillary deeply notched; no apparent teeth anywhere. Eyes rather high, partly covered both anteriorly and posteriorly by a membranaceous veil ; their diameter about one-fifth the length of the head ; more than one diameter between them and the end of the snout. Subopercle obliquely rounded off at bottom, but the curvature of the posterior margin of the opercle and subopercle taken together not very considerable. Preopercle marked with vein-like ramifications; opercle similarly veined, and also striated below, as in the species last described, though more finely. Scarce any trace of a lateral line.

Scales moderately large, thin and membranaceous. One from the middle of the side of a sub-oblong form, the hinder angles rounded, its length only two-thirds of its breadth ; the greater portion of the surface marked with exceedingly delicate striæ, scarcely visible without a strong lens, the anterior margin pectinated, and with a slightly projecting lobe in the middle. The scales as they approach the tail, become longer in proportion to their breadth, the basal margin more rounded, and sometimes with a strongly projecting lobe in the middle. The pectinations are longest on the scales covering the nape.

The dorsal commences a little behind the middle point of the oval of the body, and the ventrals are attached a little in advance of that fin. The anal commences a little behind the termination of the dorsal, and reaches to the commencement of the fleshy part of the tail : the last ray in both dorsal and anal is slightly lengthened beyond the preceding ones. The pectorals reach to the ventrals, and are contained about once and two-thirds in the length of the head. Caudal deeply forked; the lower lobe a little longer than the upper one: the base appears to have been covered with minute scales. Above the pectorals is a thin membranaceous lanceolate scale, more than half the length of the fin: a somewhat similar scale in the axillæ of the ventrals, but shorter in proportion ; another below those fins.

COLOUR.—" Body silvery : dorsal scales iridéscent with green and copper: head greenish: tail yellow."—D.

A *second specimen* agrees with the above in form, but is smaller, measuring only 7½ inches. The colours, when recent, according to Mr. Darwin's notes, were as follows :—" Scales silvery iridescent; back especially greenish; caudal fin yellow : remarkable for a circular dark green patch behind the gill-cover."—D.

Habitat, Bahia Blanca.

This species was caught by Mr. Darwin in the net, on a sandbank, at Bahia Blanca. It is well characterized by its strongly-pectinated scales, and does not appear to have been noticed by authors ; though it would seem in that respect to have some affinity with the *Clupea fimbriata* of Bowdich.*

<h3 style="text-align:center">ENGRAULIS RINGENS. *Jen.*</h3>

E. capite compresso, grandiusculo, quartam partem longitudinis totius æquante: rostro acuto, ultra maxillam superiorem mediocriter prominulo ; mandibulâ angustâ, dentibus lateralibus (ut etiam in maxillâ) minimis: corpore compresso : pinnis ventralibus infra, vix ante, initium pinnæ dorsalis exorientibus : squamâ longissim â membranaceâ super pinnam pectoralem retrorsum productâ.

D. 15 ; A. 19 ; C. 19, &c. ; P. 16 ; V. 7.

LONG. unc. 5.

FORM —Closely resembling the common Anchovy, but the head decidedly larger and longer, being one-fourth of the entire length.† Eye larger, but bearing an equal proportion to the size of the head; also rather nearer to the tip of the snout in consequence of this last not being so acute and much produced. Lower jaw rather narrower, from the greater compression of the head and body. Maxillary, and its fine serratures on the edges for teeth, similar.

The depth of the body is about one-sixth of the entire length. The dorsal commences at the middle point of the length, reckoning this last to the base of the caudal fork, and terminates a little before the commencement of the anal : the first ray is not half the length of the second and third, which equal three-fourths of the depth : the fifth and succeeding rays become gradually shorter than those which precede. The first ray in the anal is likewise very short, and scarcely one-third of the next following. The ventrals arise almost directly under the first ray of the dorsal, being scarcely at all in advance ; when laid back, they do not reach half-way to the anal. Above the pectoral is a long membranaceous scale equalling, or very nearly, the fin itself.

* *Excursions in Madeira*, p. 234, fig. 44.

† In the *E. enchrasicholus*, it is hardly one-fourth of the length, excluding caudal.

27

Colour.—Not noticed in the recent state. In spirits, it appears silvery, with the back and upper part of the sides deep dusky blue, the two colours separated by a well-defined line.

Habitat, Iquique, Peru.

This is probably an undescribed species of *Engraulis*; nor am I aware that authors have hitherto noticed any from the west coast of America. Mr. Darwin obtained two specimens which are precisely similar to each other. The species closely resembles the common European Anchovy,* differing principally in its larger head, and more backward ventrals in respect to the dorsal.

Family.—PLEURONECTIDÆ.

Mr. Darwin's collection contains individuals of five species belonging to this family, besides the drawing of a sixth; but the specimens brought home are dry, and badly preserved. Two appear to have been previously noticed; but it is difficult to pronounce upon the other three with certainty, neither do they admit of being very accurately described. These last, therefore, I shall not venture to name, but merely point out a few of their principal characters, adding the localities whence they were obtained.

The species, of which there is a drawing, I conceive to be certainly new; and as its characters are very distinguishable, I shall name it in honour of the gentleman, one of the officers of the Beagle, by whom the drawing was made.

1. Platessa Orbignyana. *Val.?*

Platessa Orbignyana, *Val.* in D'Orb. Voy. dans l'Amer. Mérid. Atl. Ichth. Pl. 16. fig. 1.

Form.—Oval; greatest breadth two and a half times in the length. Eyes on the left side, near together, and equally in advance. Teeth in a single row, sharp-pointed, moderately strong, rather widely separate: posterior extremity of the maxillary cut nearly square. Dorsal commencing in a line with the eyes, and leaving a space between it and the caudal. The lateral line takes a sweep over the pectoral. Upper or eye side of the body slightly rough, with the scales finely ciliated; under side smooth, the scales on this side not ciliated.

Colour.—"Above dirty reddish brown; beneath faint blue: iris yellow."—D.

Length 8 inches 9 lines.

Habitat, Bahia Blanca, where it is said to be plentiful.

This species agrees so well with the figure of the *P. Orbignyana* in D'Orbigny's Voyage, that I have little hesitation in considering it the same,—but as

* I am indebted to Mr. Yarrell for the loan of a specimen of our common Anchovy for comparison.

T

no description of this last has been yet published, it is still possible I may be mistaken.

2. PLATESSA——?

FORM.—Very similar to the last species, from which it scarcely seems to differ, except in having the teeth smaller, and somewhat more numerous and closer together; also in the maxillary, which is more dilated at its posterior extremity, and more obliquely truncated. The scales are extremely similar.

Length 6 inches 6 lines.

COLOUR.—Not noticed.

Habitat, King George's Sound, New Holland.

HIPPOGLOSSUS KINGII. *Jen.*

PLATE XXVI.

H. fuscus: corpore ovato, lato: oculis sinistris, haud valde approximatis: dentibus acutis, fortioribus: lineá laterali anticè arcuatá: pinná dorsali supra oculos initium capienti, dimidio anteriore humillimo, posteriore modicè elevato: ventralibus distinctis, haud anali continuis: caudali subquadratá, radiis mediis cæteris paululum longioribus.

D. 18 et 48 ; A. 51 ; C. 14 ; P. 11 ; V. 6.

FORM.—Breadth, not including the dorsal and anal fins, half the length of the oval of the body. Eyes on the left side, apparently distant from each other about two diameters; the upper one a little behind the lower. Teeth sharp and strong, forming a very regular series. The lateral line takes a sweep over the pectoral fin. The dorsal commences above the upper eye; the first half, or until it gets above the extremity of the reclined pectoral, is very little elevated, and much lower than the rest of the fin, with the membrane apparently notched between the rays; the remainder of the fin attains a moderate elevation, and there is an abrupt transition from the former to the latter portion. The anal answers to the elevated portion of the dorsal: both these fins fall short of the caudal by a small space. Pectorals short, and of a somewhat triangular form. Ventrals very distinct, free, placed right and left, with the rays a little projecting beyond the membrane; which last character appears also in the dorsal and anal. Tail somewhat square, but the middle rays slightly projecting beyond the lateral ones in the form of an obtuse lobe.

COLOUR.—Represented in the drawing of a uniform light brown.

Habitat, Valparaiso.

This is the species of which, as before stated, no specimen was brought home, but only a coloured drawing made by Mr. Phillip King, an officer of the Beagle, for Capt. FitzRoy. The drawing appears to have been done with accuracy, and from it the above description has been taken. The fin-ray formula, however, was computed from the recent fish, the above numbers being marked upon the drawing.

The teeth appear to indicate this species as belonging to *Hippoglossus* rather

2

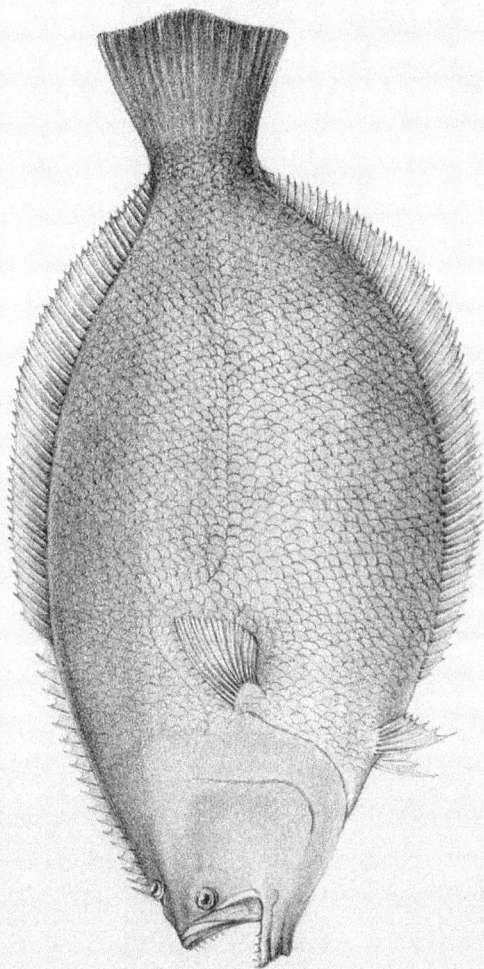

Hippoglossus Kingii

Lithog. from Nature by W. Buelow.

than to *Rhombus*, though possibly it may be found hereafter to serve as the type
of a distinct subgenus in this family. The form of the dorsal fin, if correctly
delineated, is remarkable. The size of the fish is not stated.

RHOMBUS——?

FORM.—Oval, approaching to rhomboidal. Breadth a little exceeding half the length. Eyes
on the right side, near together, equally in advance, or the lower one perhaps rather more
forward than the upper; between them a double osseous ridge. On the under side of the
head, and nearly answering in position to the upper eye, is a deepish cavity, from whence
proceeds a tentaculiform appendage four or five lines in length. Teeth very small, sharp,
in scarcely more than two rows, and apparently confined to the under side. Lateral line
sweeping over the pectoral. Dorsal commencing above the upper lip, and reaching nearly
to the caudal, but leaving a minute space. Both sides of the body are smooth, but the upper
one appears to have lost its scales. Pectoral on the eye side about three-fourths the length
of the head.

Length 5 inches.

COLOUR.—" Above pale purplish brown, with rounded darker markings."—D.

Habitat, Bahia Blanca, Coast of Patagonia.

ACHIRUS LINEATUS. *D'Orbig.*

Achirus lineatus, *D'Orb.* Voy. dans L'Amer. Mérid. Atl. Ichth. Pl. 16. fig. 2.

FORM.—Body oval, but with the dorsal and anal fins included, approaching orbicular; the greatest
breadth rather more than half the length. Eyes on the right side, moderately near together,
the upper one a very little in advance. Lower jaw longest, projecting beyond the snout.
Teeth forming a velutine band, very minute, and scarcely sensible except to the touch,
confined to the side opposed to the eyes. Preopercle distinct from the opercle. A few short
thread-like cirri on the under side of the head; two at the extremity of the snout being
rather longer and more conspicuous than the others. Lateral line nearly straight throughout
its course, somewhat higher at its commencement than afterwards, but taking no sweep. Both
sides of the body rough, with ciliated scales, but the upper one most so. The dorsal commences above the upper lip, and reaches, as also the anal, almost quite to the caudal: this
last rounded. Pectorals entirely wanting.

Length 9 inches.

COLOUR.—Not noticed.

Habitat, Rio Plata.

This species was obtained by Mr. Darwin in the market at Buenos Ayres,
where it is said to be eaten. It so exactly accords in form with the figure of the
A. lineatus in D'Orbigny's Voyage, that I have little hesitation in considering it
the same, though, from the specimen being dried, there are no vestiges left of the
transverse lines. Whether it be the *A. lineatus* of any other author I am uncer-

tain. It approaches, however, very closely the *Passer lineis transversis notatus* of Sloane.*

PLAGUSIA——?

FORM.—Oval, but narrow, and much elongated for a *Sole*, the breadth in the middle being three and a half times in the length. Eyes on the left side, very small, and closely approximating, equally in advance, or if any difference, the lower one a little first. Mouth small, with velutine teeth on the supine side, but apparently none on the upper: snout a little produced in a point beyond it. The dorsal and anal unite with the caudal, which terminates in rather a fine point. No trace of any pectorals above or below. Scales strongly ciliated, especially above, and both sides of the body rough.

Length 7 inches.

COLOUR.—Not noticed.

Habitat, San Blas, Coast of Patagonia.

This species is very nearly allied to the *Plagusia Braziliensis* of Spix's work,† but it appears to differ in having the eyes one over the other, or the lower one perhaps a little in advance, instead of the upper one a little before the lower.

FAMILY.—CYCLOPTERIDÆ.

1. GOBIESOX MARMORATUS. *Jen.*

PLATE XXVII. Fig. 1.

G. dorso et lateribus pallidè fuscis, nigro reticulatis et fasciatis: dentibus anterioribus majoribus, in maxillá superiore subconicis, in inferiore incisivis: operculo posticè mucrone obtuso armato: membraná branchiali spiná gracili, subduplici, (præter radios solitos,) instructá, magná ex parte celatá, apice exserto: pinná dorsali tredecim-radiatá.

B. 6; D. 13; A. 11; C. 14 vel 15; P. 20 vel 21.

LONG. unc. 2. lin. 7.

FORM.—Head very large, broad and much depressed, with the snout rounded nearly in an exact semicircle. Body compressed behind, and suddenly tapering behind the pectorals. The length and breadth of the head are equal, each being one-third of the entire length, excluding caudal. Gape wide, reaching nearly to beneath the anterior angle of the eye. Teeth strong, and somewhat crowded in front; in the upper jaw bluntly conical, or slightly curved, but of irregular size, with minuter ones behind; very small at the sides of the jaw, and apparently here but in a single row: below, the six middle teeth are incisor-like, and project forwards; on each side of these are two or three similar to those in front above, then follow some minuter ones at the sides. Eyes rather more than a diameter apart. Gill-open-

* Nat. Hist. of Jamaica, Pl. 246, fig. 2.
† Pisces Brazil. p. 89, tab. L.

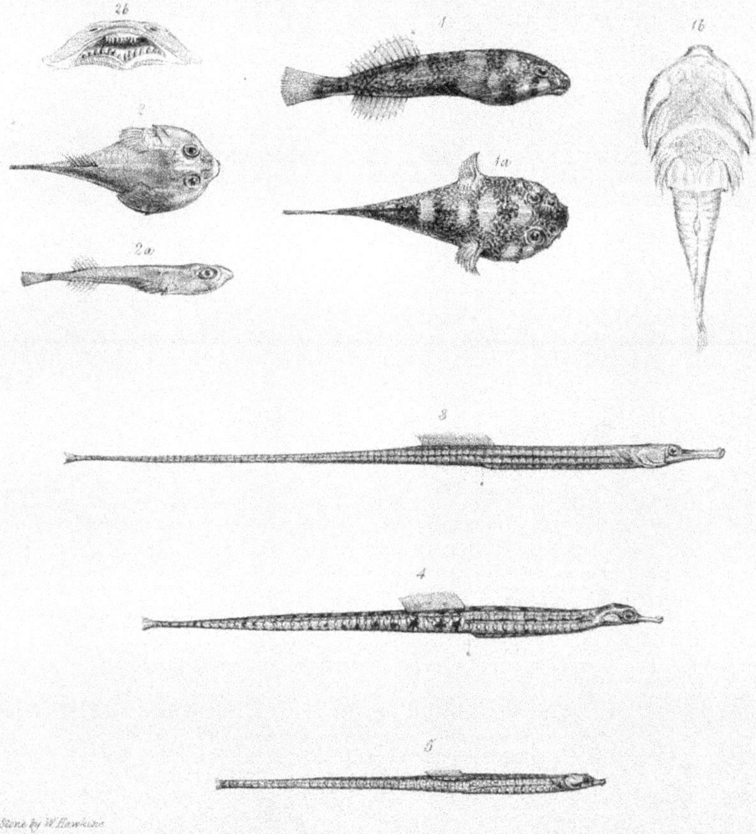

From Nature on Stone by W. Hawkins

1 *Gobiesox marmoratus* — 1a " " Dorsal View — 1b " " Under Side

2 *Gobiesox pœcilophthalmos.* — 2a " " Lateral View — 2b " " Magnified Row of Teeth.

3. *Syngnathus acicularis.* — 4 " *conspicillatus.* — 5 " *crinitus.*

All Nat. Size

ing wide, the membrane free all round, with six rays. Opercle terminating behind in a blunt point: there is also a kind of double spine concealed in the thickness of the branchial membrane, in front of the ordinary branchial rays, the extreme end of one portion of which projects a little beyond the margin.

The dorsal commences a little beyond the middle of the entire length, and leaves a space between it and the caudal; the rays nearly equal, except the first, which is short. The anal begins under the fourth or fifth dorsal ray, and extends a trifle further than that fin. Caudal slightly rounded.

COLOUR—(*In spirits.*) Back and sides light brown, reticulated with black: the reticulations have a tendency to form three or four broad fasciæ across the back. Under parts yellowish.

Habitat, Archipelago of Chiloe.

This and the following species appear to belong to the genus *Gobiesox* of Cuvier's " Regne Animal," and are probably new. Two specimens of the one above described were found by Mr. Darwin under stones off the island of Lemuy, in the Archipelago of Chiloe.

2. GOBIESOX PŒCILOPHTHALMOS. *Jen.*

PLATE XXVII. Fig. 2.

G. fuscescenti-albid is, immaculatus: dentibus anterioribus majoribus, supra et subtus incisivis : operculo posticè spinâ acutâ armato; membranâ operculari margine, supra spinam, cirris paucis filamentosis fimbriato; membranâ branchiali spinâ nullâ : pinnis dorsali et anali septem-radiatis.

B. 6; D. 7 ; A. 7; C. 12 ; P. 23 ;

LONG. unc. 1. lin. 10.

FORM.—General form the same as that of the last species, including the proportion of head to body. Snout equally rounded. Teeth on the whole similar, but the upper ones in front, as well as the lower, incisor-like. Eyes rather larger, closer together, less than a diameter apart. Differs essentially from the *G. marmoratus* in the form of the opercular spine, which is much sharper, as well as somewhat longer and slenderer; also in having no spine concealed in the branchial membrane : the lower front part of the opercular membrane, just above the spine, is fringed with a few thread-like filaments. The number of branchial rays is the same.

The dorsal and anal are both shorter, and appear to have only seven rays each : the anal reaches a little nearer the caudal. The pectorals on the contrary have rather more rays.

COLOUR—(*In spirits.*) Every where of a uniform very pale brown, or brownish white, without any markings whatever. The eyes were probably very brilliant in the living fish, the irides still showing traces of what seems to have been blue and golden pink.

Habitat, Galapagos Archipelago.

A single individual of this species was obtained by Mr. Darwin in tidal pools at Chatham Island, in the Galapagos Archipelago.

FAMILY.—ECHENEIDIDÆ.

ECHENEIS REMORA. *Linn.*

Mr. Darwin took a small specimen of this fish from off a shark in the Atlantic Ocean, near St. Paul's Rocks. It is not four inches long. It has eighteen pairs of laminæ on the head; and a rough disk on the middle of the tongue:* caudal lunate.

FAMILY.—ANGUILLIDÆ.

ANGUILLA AUSTRALIS. *Richards.*

Anguilla australis, *Richardson*, Proceed. of Zool. Soc. 1841, p. 22.

FORM.—Very similar to the *A. latirostris*, Yarr., but the upper jaw rather shorter and broader, making the gape, which reaches to a vertical line from the posterior part of the orbit, wider. Teeth rather stronger. Dorsal commencing considerably beyond the first third, and not much in advance of the middle point, of the entire length; much less elevated than in the *A. latirostris*, its height scarcely exceeding one-fifth of the depth, which last is about one-seventeenth of the entire length. Vent a little posterior to the commencement of the dorsal.

The distance from the end of the snout to the insertion of the pectorals is rather less than one-eighth of the entire length: the form of the pectorals is lanceolate. The tail is rounded, much as in the *A. latirostris*.

	in.	lin.
Length (entire)	17	3
From end of snout to commencement of dorsal	7	6
From the same to insertion of pectoral	2	2
From the same to vent	7	9

COLOUR—(*In spirits.*) Appears similar to that of the common eel.

Habitat, New Zealand.

The above eel was procured by Mr. Darwin in fresh water in the month of December, in the Bay of Islands, New Zealand. It so nearly accords with the *A. australis* of Dr. Richardson from Van Dieman's Land, that I can hardly suppose it to be a distinct species. The vent, however, would seem to be a trifle backwarder, and the body deeper in proportion to its length. Without seeing more specimens, it is impossible to say what importance is to be attached to these points of discrepancy.

* I notice this circumstance, because Mr. Lowe, in the "Proceedings of the Zoological Society," (1839, p. 89.) has briefly described two species of this fish, which he calls *E. Remora* and *E. pallida* respectively, the former having the tongue *smooth*, and the latter *rough* in the middle, besides other differences.

The above specimen obtained by Mr. Darwin, as well as two others in the Museum of the Cambridge Philosophical Society, have the tongue rough; though in their other characters, especially colour, they would seem to be Mr. Lowe's *Remora*. Cuvier, in his "Regne Animal," appears to consider the rough tongue as characteristic of the whole genus.

Conger punctus. *Jen.*

C. lateribus fasciis transversis fuscescenti-rubris, interstitiis angustis griseis : rostro brevi, obtuso ; maxillis subæqualibus : pinnâ dorsali initium supra pectoralem capienti : cute corporis puncturis parvis creberrimè aggregatis impressâ.

FORM.—Body much compressed, except at the anterior extremity. Depth less than one-eleventh of the entire length. Head contained about seven and a half times in the same. Snout short and rounded. Jaws nearly equal, the upper scarcely longer than the lower. Gape scarcely reaching beyond a vertical from the anterior part of the eye. Teeth velutine. A row of very conspicuous pores round the edges of both jaws. The whole body, but not the head, thickly studded all over with small pores, much crowded, and appearing like pin-holes.

The pectorals are rather more than half the length of the head. The dorsal commences immediately above them, and has a moderate elevation of about one-third of the depth. The vent is a little posterior to the termination of the first third of the length, and the anal is immediately behind it. The dorsal and anal unite to form a moderately pointed caudal.

Length 3 inc. 3 lines.

COLOUR—(*In spirits.*) Sides very regularly banded with fourteen or fifteen transverse reddish brown fasciæ : the fasciæ extend on to the dorsal fin, and are much broader than the intervening spaces. All the under part of the head, belly as far as the vent, an irregular patch on the cheeks, and the spaces between the bands on the sides, yellowish.

Habitat, Tierra del Fuego.

This appears to be a new species. The individual described above is quite small, and stated in Mr. Darwin's notes to be the young of another and larger specimen which he also captured, but of which he does not mention the exact size, and which unfortunately does not appear in his collection. He has, however, mentioned the colours, which appear similar to those given above, and are as follows :—" Sides with transverse bars of chocolate and brownish-red, separated by narrow grey spaces." Whether the form and proportions of the adult agree exactly with those of the young as above detailed must be left for future observers to determine. The colours, however, appear well to characterize the species, aided by the minute punctures with which the whole body is covered.

This species was taken by Mr. Darwin at the roots of fucus, at the east entrance of Beagle Channel, Tierra del Fuego. The larger specimen is said to have been very active.

1. Muræna lentiginosa. *Jen.*

M. purpurascenti-fusca ; maculis circularibus, parvis, flavis : capite et rostro valde compressis ; fronte declivi : maxillis subelongatis, angustis, æqualibus, acutis ; dentibus acutis, in maxillâ superiore anticis uni- lateralibus bi-seriatis ; in inferiore anticis bi- lateralibus uni-seriatis ; in vomere uni-seriatis ; anticis supra subtusque, lateralibus secundariis supra, et vomerinis, fortibus : pinnâ dorsali anticè obsoletâ.

LONG. unc. 20. lin. 6.

FORM.—Very much compressed about the head and jaws. Body tapering posteriorly; the depth in the middle equalling about one-thirteenth of the entire length. Head, measured to the branchial orifice, about one-seventh. Profile falling obliquely in a straight line from the nape to the extremity of the snout. Jaws very narrow, rather lengthened and sharp-pointed, equal. Gape deeply cleft, reaching as far back behind the eyes as it advances before them. Teeth compressed at the sides, very sharp, slightly hooked and pointing backwards; above, in a single row in front, in two rows at the sides; below, in two rows in front, and in a single row at the sides; in each case, however, the secondary row is very imperfect, some of them appearing to have been lost; also a row down the vomer, but interrupted in the middle of the series: the front teeth above and below, and the secondary ones at the sides of the upper jaw, are much stronger than the others; but the first three on the vomer, being those anterior to the blank space, are perhaps longer and more developed than any in the jaws. Two tubular orifices above the eyes, and two at the extremity of the snout. Eyes distant from the end of the snout twice their own diameter. Branchial orifice of the same size as the eyes. Three or four large pores arranged in a line along the edge of the upper jaw, but none apparent on the lower.

Dorsal fin thick and fleshy, and not very distinguishable from the body, excepting posteriorly, so that its exact point of commencement cannot be fixed with precision. Vent a trifle in advance of the middle point of the entire length. Anal fin still less distinguishable than the dorsal.

COLOUR.—" Fine dark purplish brown, with yellow circular spots."—D.—The spots are mostly small, and many of them not bigger than large pin's heads. They are smaller and more crowded about the head than elsewhere, giving a freckled appearance.

A *second specimen* is smaller than the above, measuring thirteen inches and a half in length. This specimen has the teeth more perfect. In the upper jaw, there is first an outer row reaching all round, in which the teeth are mostly small and regular, but towards the front mixed with some much longer ones; behind this, about the middle of the sides, is a short secondary row consisting of five or six teeth as long as those in front in the first row : in the lower jaw, the secondary row consists likewise only of four or five long teeth, but here they are placed in front instead of at the sides. Mr. Darwin's notes respecting the colours of this smaller specimen are as follows: " Dark reddish-purple brown, with pale, or whitish-brown spots: eyes bluish."

Habitat, Galapagos Archipelago.

The larger of the two specimens above described was taken by Mr. Darwin at Charles Island, the smaller one in tidal pools at Chatham Island, in the Galapagos Archipelago. It appears to be an undescribed species, though bearing much similarity to the *M. Meleagris* of Shaw.

2. MURÆNA OCELLATA.

Gymnothorax ocellatus, *Spix et Agass.* Pisces Brazil. p. 91. tab. 50 b.

FORM.—Head but moderately compressed. Snout rather short and blunt. Jaws equal. Gape reaching a little beyond the posterior part of the orbit. Teeth apparently in only a single row above and below, very strong and sharp at the extremity of the jaws : none at the anterior part of the vomer, but a few very short ones not easily seen at the back part of the median line of the palate. Two tubular orifices at the extremity of the snout, but above the eyes only two simple pores not prolonged into tubes. Eyes rather large, much exceeding in size the branchial orifice ; scarcely more than one diameter between them and the end of the snout. Two or three large pores along the edges of both jaws Dorsal very distinct, commencing above the branchial orifice. Vent a little before the middle. Tail gradually tapering to a rather fine point.

Length 12 inc. 9 lines.

COLOUR.—(*In spirits.*) Head and trunk brown, with round whitish spots. Dorsal and anal spotted with black and white, the black spots occupying the edge of the fin. Extremity of the tail imperfectly banded with white and dusky brown. Belly pale.

Habitat, Rio de Janeiro.

This elegant and well-marked species, first discovered by Spix on the Brazilian coast, was taken by Mr. Darwin in the harbour of Rio de Janeiro.

3. MURÆNA —— ?

FORM.—Head moderately compressed, rising considerably at the nape. Body slender, somewhat ensiform behind, and tapering towards the tail. Snout of moderate length. Jaws equal, or the upper one perhaps a very little longer than the lower. Gape reaching as far behind the eye, as it advances before it. Teeth partially in two rows above, in one below ; sharp and strong at the extremity of the jaws, and on the anterior part of the vomer. Two tubular orifices at the extremity of the snout, but only simple pores above the eyes. Three or four large pores along the edges of the upper and under jaws. Eyes distant one diameter and a half from the end of the snout. Dorsal distinct, commencing almost on the occiput, and in advance of the branchial orifice. Vent before the middle. Anal commencing a little behind it, and, like the dorsal, distinct, but rather less so.

	in.	lin.
Length	10	0
Depth, fins not included	0	6
From end of snout to branchial orifice . .	1	4
From the same to vent	4	3

COLOUR.—(*In spirits.*) Rather dark brown, nearly uniform, but here and there with lighter mottlings. The lower jaw appears to have had a row of whitish spots encircling the pores.

The species of *Muræna* above described was taken by Mr. Darwin at Porto Praya, Cape de Verds. The individual being small, and possibly not having attained its permanent characters, I have forborne giving it any name, though I have not been able to identify it in the works of authors.

U

4. Muræna ——?

Form.—Snout rather compressed before the eyes, not very long, and slightly obtuse. Upper jaw a
   very little in advance of the lower. The gape extends behind the eyes, but the posterior
   portion is not equal to the anterior. The teeth, tubular orifices, and pores, are very much the
   same as in the species last noticed. Dorsal very distinct, commencing in advance of the
   branchial orifice. Anal not so distinct as the dorsal.

Length 5 inc. 6 lin.

Colour.—Brown, but with some lighter specks and mottlings, more particularly on the lower jaw
   and on the fins.

Taken by Mr. Darwin at Tahiti.   Probably a new species, but, as in the last
case, the specimen is young and not easily determinable.

LOPHOBRANCHII.

FAMILY.—SYNGNATHIDÆ.

1. SYNGNATHUS ACICULARIS. *Jen.*

PLATE XXVII. fig. 3.

S. flavo-brunneus : corpore gracillimo, compresso, heptagono ; caudá quadrangulá : vertice plano ; cristá occipitali parum conspicuá ; rostro longo, compresso, verticaliter capite angustiore, margine superiore acuto prope recto: pinná dorsali totá multum ante medium longitudinis sitá ; pinnis pectoralibus parvis, anali minutissimá, caudali distinctá.

LONG. unc. 5. lin. 10.

FORM.—Very similar to the *S. Acus*, but the body rather more compressed. The angles are the same, and the middle lateral ridges of the trunk rise upwards in a similar manner to terminate behind the dorsal fin. There are about seventy transverse shields or plates in the whole length, eighteen of which lie between the gills and the vent. Head much compressed about the gills, contained with the snout about eight and a half times in the entire length. Crown nearly flat, with very little of an occipital ridge; profile falling obliquely, but not much out of a straight line; between the eyes a slight hollow. Snout elongated, a trifle more than half the entire length of the head, compressed, the upper edge sharp and nearly horizontal in front of the nostrils, vertically much narrower than the head.

The dorsal commences at one-third of the entire length, and occupies a space about one-tenth of the same, terminating before the middle: the number of rays is about forty or more. Vent about underneath the seventh dorsal ray. Anal extremely minute, of only one or two rays. Pectorals very small. Caudal distinct, much as in *S. Acus*.

COLOUR.—(*In spirits.*) Of a nearly uniform yellowish brown, paler underneath.

Habitat, Valparaiso.

This species, taken by Mr. Darwin at Valparaiso, would seem to represent in that quarter of the globe the *S. Acus* of the European seas, which, on the whole, it much resembles, though there are several slight differences on a close comparison. It is a female specimen, being without the abdominal pouch, and is probably not full-sized. The dorsal fin being a little injured, and the rays very delicate as well as close-set, it is hardly possible to tell the exact number. The anal exists, but it is so extremely minute that it might easily be overlooked.

2. SYNGNATHUS CONSPICILLATUS. *Jen.*

PLATE XXVII. fig. 4.

S. grisens, fasciis transversis fuscis ; genis albicantibus, vittis duábus angustis longitudinalibus nigro-fuscis: corpore crassiore, subcylindrico, hexagono ; caudá quadrangulá : vertice elevato ; cristis occipitali et nuchali distinctis: oculis magnis pro-

*minulis : fronte declivi, in descensu sinuato : rostro brevi, gracillimo, subcylindrico :
pinná dorsali paulo ante medium longitudinis desinenti : ano infra radium primum
dorsalem sito : pinnis pectoralibus parvis, unuli minutissimá ; caudali distinctá.*

D. 31 ; A. 3? ; C. 10 ; P. 14.

Long. unc. 4. lin. 7.

Form.—Body rather thick and somewhat cylindrical; the greatest depth and thickness nearly equal,
the former being about one-twenty-seventh of the entire length From the head to the vent
hexagonal, the middle lateral ridges terminating abruptly, when opposite the commencement of
the dorsal fin, without inclining either upwards or downwards. Fifteen transverse plates
between the gills and the dorsal fin : only fifty-four in all, the tail not tapering so much as in
many other species. Head much pinched in at the gills, but rather full and protuberant about
the cheeks : its length ten and a half times in the entire length. Crown high and convex: a
ridge commencing at the occiput passes backward to the nape. Eyes large and full, with
somewhat of a spectacled appearance; their diameter equal to the whole depth of that part of
the head ; the orbits rising in ridges above them, with the intervening space concave. From
between the eyes the profile descends in a sinuous curve to the base of the snout, which
is short, slender, very narrow, and almost cylindrical. The length of the snout is less than half
the entire length of the head ; its breadth, vertically, only one-third the depth of the same
taken behind the eyes.

The dorsal commences beyond one-third of the entire length, occupies one-ninth of the
same, and terminates a little before the middle : nearly even, and rather high, more than
equalling the depth of the body underneath. Vent about underneath the first ray, but almost
in advance of the dorsal fin altogether. Anal extremely minute. Pectorals very small.
Caudal rays distinct.

Colour.—Trunk greyish-brown, with deep brown interrupted transverse fasciæ. In front of the
dorsal, the fasciæ terminate at the middle lateral ridge, below which the sides are spotted.
Dorsal fin also a little spotted. Cheeks whitish, with two very distinct narrow longitudinal
vittæ extending backwards from the eyes to the posterior part of the opercle.

Habitat, Tahiti.

A well-marked species, and apparently undescribed. The only specimen in
the collection is a female, and, like the last, perhaps not full-sized.

3. SYNGNATHUS CRINITUS. *Jen.*
PLATE XXVII. fig. 5.

*S. griseus ; ventre, et maculá operculari, nigricantibus : corpore crassiore, antice
heptagono, postice quadrangulo, angulis acutis : vertice parum elevato ; cristis occi-
pituli et nuchali distinctis : rostro brevissimo, subcylindrico, capite angustiore, postice supra carinato, apice subrecurvo : cirris duobus, minutis, filamentosis, palpebralibus : pinná dorsali paulo ante medium longitudinis desinente ; ano infra initium
ejus sito : pinnis pectoralibus et caudali parvis ; anali nullá.*

Long. unc. 3. lin. 5.

FORM.—Body thickish, the greatest depth and thickness nearly equal, the former about one-twenty-fifth of the whole length. From the head to the vent heptangular; tail quadrangular: all the angles sharp and distinctly marked. The middle lateral ridges in the heptangular portion pass downwards at their extremities to terminate at the vent. Sixteen transverse plates before the dorsal: only fifty-two in the whole length. Head short, about one-eleventh of the entire length, not more compressed than the body. Crown not much elevated, but with distinct occipital and nuchal ridges. Orbits rising in ridges above the eyes, the interocular space being hollowed out: also a ridge commencing between the eyes, and passing forwards along the base of the snout, but not reaching to its extremity. Snout itself very short, its length only one-third the entire length of the head, narrower than the head vertically, nearly cylindrical, the tip slightly recurved. A few very short minute filamentous threads scattered about the head, more particularly one over each eye.

Dorsal placed much as in the last species, and terminating a little before the middle ; the rays delicate and not easily counted, about twenty. Vent beneath the commencement of the dorsal. No anal distinguishable even under a lens. Pectorals very small. Caudal moderately distinct.

COLOUR.—Grey : a spot on the gill-cover, and the belly, dusky. The carinæ which form the edges of the under surface of the body are darker still, and shew a fine dark line on each side extending to the caudal.

Habitat, Bahia Blanca, Northern Patagonia.

Apparently another new species of this genus, taken by Mr. Darwin at Bahia, and, like the last, well-marked ; especially by the short filaments above the eyes, which I am not aware occur in any other known species.

PLECTOGNATHI.

Family.—TETRODONTIDÆ.

1. Diodon nycthemerus. *Cuv.*

Diodon nycthemerus, *Cuv.* Mém. du Mus. tom. iv. p. 135. pl. 7.

A species of *Diodon* in Mr. Darwin's collection, the number attached to which has been lost, and of which the locality is in consequence unknown, appears referable to the *D. nycthemerus* of Cuvier.

The spines are long, measuring three quarters of an inch in length; round, sharp, and not very close together. There are five in the front row between the eyes, seven in a transverse row between the pectorals, and ten or eleven between the snout and the dorsal in a longitudinal one: none exactly on the upper part of the tail, but one on each side of the base of it, a little below the termination of the dorsal fin, and a corresponding pair still lower down. The spines on the belly are shorter, and rather closer together than those on the back. One of those on the back in this specimen is accidentally forked.

The true teeth appear on the surface of the jaws like minute scales, as in several species of the genus *Scarus*.

The fin-ray formula is as follows:

D. 13 ; A. 13 ; C. 9 ; P. 20.

Length 5 inches 6 lines.

The colours, so far as can be judged, the specimen being in spirits and not in very good condition, answer to Cuvier's description of them with tolerable exactness.

2. Diodon rivulatus. *Cuv.*

Diodon rivulatus, *Cuv.* Mém. du Mus. tom. iv. p. 129. pl. 6.

An individual apparently of this species was picked up by Mr. Darwin on the shore of the Rio Plata at Maldonado. It agrees with Cuvier's description, excepting that the undulating lines are not visible, probably owing to the state of the specimen when found.

The spines are short, barely a quarter of an inch in length, but very strong, compressed, and resembling canine teeth. There are three in the first row between the eyes; about six in a transverse row across the back, and seven or eight in a longitudinal one. Beneath they are shorter and more numerous. The orbits are elevated in ridges, and project forwards over the eyes. Two very small barbules attached to the lower lip. Surface of the jaws smooth, the teeth not appearing as scales.

D. 11 ; A. 10 ; C. 8 ; P. 22.

Length 5 inc. 3 lin.

As Cuvier observes, the *D. geometricus* of Bl. and Schneid.* approaches very
closely this species, and 1 can hardly think it to be distinct. Yet neither in Mr.
Darwin's specimen, which in all other respects agrees exactly with Schneider's
figure, do I discern any appearance of the hexagonal meshes on the surface of
the body.

3. DIODON ANTENNATUS. *Cuv.?*

Diodon antennatus, *Cuv.* Mém. du Mus. tom. iv. p. 131. pl. 7.

A third species of *Diodon*, brought home by Mr. Darwin, and taken by him at
Bahia, in Brazil, is either the young of the *D. antennatus* of Cuvier, or else new;
but the only individual in the collection is quite small, and not more than an inch in
length, excluding caudal. The fleshy filaments above the eyes, which, according
to Cuvier, so peculiarly distinguish the *D. antennatus*, are very distinct,—but I
see none on the sides. The ground colour would seem darker than he describes,
so as to render the spots and markings on the upper parts not distinguishable
from it now, if they ever existed. In spirits it appears of a nearly uniform deep
brown red. The spines, or rather papillæ, are also shorter than represented in his
figure; but this may be only the effect of immaturity.

According to Mr. Darwin, the colours when recent were as follows:—" Above
blackish brown, beneath spotted with yellow. Eye with the pupil dark blue; iris
yellow, mottled with black." It is added:—" On the head four soft projections;
the upper ones longer, like the feelers of a snail."

Mr. Darwin observes, " that the dorsal, caudal, and anal fins, in this species,
are so close together that they act as one: these, as well as the pectorals, are in a
continued tremulous motion even when the fish is otherwise motionless. The
animal propels its body by using the posterior fins in the same manner as a boat
is sculled, that is, by moving them rapidly from side to side with an oblique surface exposed to the water. The pectoral fins have great play, which is necessary
to enable the animal to swim with its back downwards."

Mr. Darwin made some further observations on the habits of this species,
which have already appeared in his " Journal," to which I may refer the reader.†
The tendency of them is to explain the process by which the water and air are
absorbed, when the *Diodon* distends itself into a spherical form; and to show that
the fish *can* swim, when floating in this state with its back downwards, which
Cuvier doubted. He thinks that the water is taken in partly for the sake of
regulating its specific gravity. He also notices a curious circumstance with
respect to this species, viz., " that it emitted from the skin of its belly, when
handled, a most beautiful carmine red and fibrous secretion, which permanently
stained ivory and paper."

* *Syst. Ichth.* pl. 96.   † pp. 13, 14.

1. TETRODON AEROSTATICUS. *Jen.*

T. capite, dorso, lateribus, et pinnâ caudali, nigro-maculatis; ventre turgidissimo, fasciis obliquis nigris: corpore undique muricato, caudâ solum exceptâ: capite brevi; fronte inter oculos paululum depresso: maxillis æqualibus: lineâ laterali nullâ: pinnâ dorsali omnino ante analem positâ: pinnâ caudali subrotundatâ.

D. 11 ; A. 10; C. 10 ; P. 11.

LONG. unc. 2. lin. 6.

FORM.—Head short. Body approaching to globular, with the skin of the belly extremely loose and capable of great inflation ; every where beset with minute prickly asperities, the extreme end of the tail alone excepted. Crown nearly flat, very slightly depressed between the eyes. Jaws equally advanced. Nostrils tubular. No appearance of any lateral line. Dorsal entirely in advance of the anal : both these fins small. Caudal slightly rounded.

COLOUR.—(*In spirits.*) Head, back, and sides to the depth of the pectorals, greyish brown, spotted with black; the spots very small and crowded on the back, but becoming larger on the flanks and tail. Belly white, with deep black oblique broad bands, inosculating in some places, so as to form large meshes. Dorsal, anal, and pectorals, plain; but the caudal very elegantly and distinctly spotted.

The ticket attached to this specimen has been lost, and its locality is in consequence unknown. In general appearance, it very much resembles the *T. lineatus* of Bloch, of which it may possibly be a variety ; but it would seem to differ from that species, in having the forehead less elevated ; in wanting the lateral line altogether, of which I can discover no trace ; and in having the whole back and upper part of the sides spotted, and not merely the tail and its fin, as is represented in the *T. lineatus.*

2. TETRODON IMPLUTUS. *Jen.*

T. sordidè metallico-olivaceus, maculis circularibus albis; ventre albo, lineis olivaceis longitudinalibus, haud admodum turgido: corpore suboblongo, magnâ ex parte lævissimo, ventre solum muricato: maxillis subæqualibus: naribus tubulosis, bifurcatis: lineâ laterali distinctâ, parum tortuosâ: pinnâ dorsali anali paulo anteriore: pinnâ caudali æquali.

D. 10 ; A. 10 ; C. 11 ; P. 16.

LONG. unc. 4. lin. 9.

FORM.—Approaching to oblong, the belly a little ventricose. Head not so short as in the last species, nor yet much produced. Body every where smooth, excepting the middle of the abdomen from beneath the pectorals to the vent, and not very prickly here. Top of the head slightly depressed between the eyes. Jaws nearly equal; the upper one, if any thing, a very little in advance. Nostrils tubular, the tubes forked from the bottom into two equal branches.

The lateral line, which is very distinct, commences behind the mouth, whence it passes under and partly encircles the eye, then arches upwards, making a long sweep, and not descending till it gets above the anal, whence it proceeds nearly along the middle towards the caudal, but loses itself before attaining to that fin. Dorsal fin rather in advance of the anal. Caudal square.

COLOUR.—"Dirty metallic olive-green, with white circular spots; belly white, with streaks of the same colour as the back."—D. The spots extend on to the basal half of the caudal, but are smaller here than on the body. A white annulus encircles each eye, and a similar one is described round the base of each pectoral. The abdominal streaks run very exactly parallel with the axis of the body, not obliquely as in the last species.

Habitat, Keeling Islands, Indian Ocean.

I can find no species noticed by authors exactly corresponding with the one described above, which was obtained by Mr. Darwin at the Keeling Islands. The form is similar to that of the *T. Honckenii* of Rüppell,* but the colours appear different. On the other hand, the markings resemble those of the *T. testudineus* of Bloch, but that species is rough all over.

3. TETRODON ANNULATUS. *Jen.*

T. dorso et lateribus nigro-fuscis, maculis circularibus atris; infra niveus: corpore oblongo, haud admodum ventricoso, ubique sed parcè muricato, rostro et caudâ exceptis: capite grandiusculo, spatio interoculari lato, parum depresso: maxillis subæqualibus: naribus cylindraceis, recumbentibus, aperturis duabus lateralibus: lineâ laterali in capite tortuosissimâ: pinnâ dorsali vix anali anteriore: pinnâ caudali æquali.

D. 8; A. 7; C. 9, &c.; P. 15.

LONG. unc. 9.

FORM.—Oblong: head rather large; the snout a little more produced than in the last species. Moderately ventricose, and apparently capable of a certain degree of inflation. No where perfectly smooth, except on the snout, tail, and here and there on the flanks; nor very rough, the prickles being minute and rather scattered, most apparent on the back, nape, (whence they advance to quite between the eyes,) and the middle of the abdomen. The interocular space is broad, equalling two and a half diameters of the eye at least, and a little hollowed out. Jaws nearly equal, the upper one perhaps a very little in advance. Nostril in the form of a small recumbent cylinder, with an opening at each extremity. Dorsal very little in advance of the anal; the first ray in each of these fins very short. Caudal square.

The lateral line is very tortuous, especially about the head. It commences at the bottom of the gill-cover, whence it ascends vertically behind the eye towards the crown, then passes over the eye towards the snout, descends again beneath the nostril to form a great loop in front of the eye, almost reaching to the corners of the mouth, whence it returns beneath the eye,

* Surely this cannot be the same as the *T. Honckenii* of Bloch?

x

and, crossing its former course nearly at right angles, proceeds along the upper part of the side, getting lower as it approaches beneath the dorsal, to terminate at the caudal.   There are also two short transverse lines ; one across the snout, connecting the loops; another across the nape, connecting the two main lines after they have assumed the usual direction.

Colour.—"Beneath snow white.   Above dark brownish-black, this colour forming a series of broad oval rings, one within another ; the outer and largest ring includes nearly the entire surface of the back and sides.   The upper surface is, in addition, marked with round spots of a darker shade.   Pectoral and dorsal fins yellowish brown.   Iris, inner edge clouded with orange ; pupil dark green-blue."—D.—In its present state, there is no indication of the rings noticed above.   The spots, which are small, and cover nearly the whole head, back, and sides, appear also sparingly on the basal half of the caudal, but not on any of the other fins.

Habitat, Galapagos Archipelago.

This species was taken by Mr. Darwin at Chatham Island, in the Galapagos Archipelago.   He observes in his notes that it makes a loud grating noise.   It is remarkable for the great tortuosity of the lateral line.   The form of the nostrils is also rather peculiar.

4. Tetrodon angusticeps. *Jen.*
Plate XXVIII.

T. supra obscure viridis: capite oblongo, subcompresso, spatio interoculari multum contracto : corpore infra ventricoso, ubique lævissimo, duobus, in summo dorso, cirris cutaneis parvis adornato : maxillis subæqualibus: naribus tubulosis, indivisis, aperturis duabus lateralibus : lineâ laterali in capite tortuosissimâ : pinnâ dorsali omnino ante pinnam analem ; caudali æquali.

D. 8 ; A. 7 ; C. 9; P. 15.
Long. unc. 9. lin. 3.

Form.—Rather more elongated than the last species ; especially in regard to the head, which is also more compressed upwards, reducing the space between the eyes to a narrow channel, much hollowed out, and not exceeding one diameter of the eye.   Body inflatable, every where quite smooth.   Jaws nearly equal, the upper one perhaps a very little in advance.   Nostrils tubular, with two lateral apertures, somewhat similar to those of the last species, but more elevated.   Lateral line similar, taking the same windings on the head.   A little behind the transverse line on the nape, and nearly above the attachment of the pectoral, are two small skinny appendages : there is also a very minute one on each side of the tail, but none elsewhere.   Dorsal wholly before the anal.   Caudal square.

Colour.—" Above dull green: base of the pectorals and dorsal black ; a white patch beneath the pectorals."—D.—The colours must have very much altered from the action of the spirit, as it now appears of a nearly uniform reddish brown, only paler beneath.

Habitat, Galapagos Archipelago.

Another apparently undescribed species of this genus, taken by Mr. Darwin at the same place as the last.   He observes in his notes that it is inflatable.

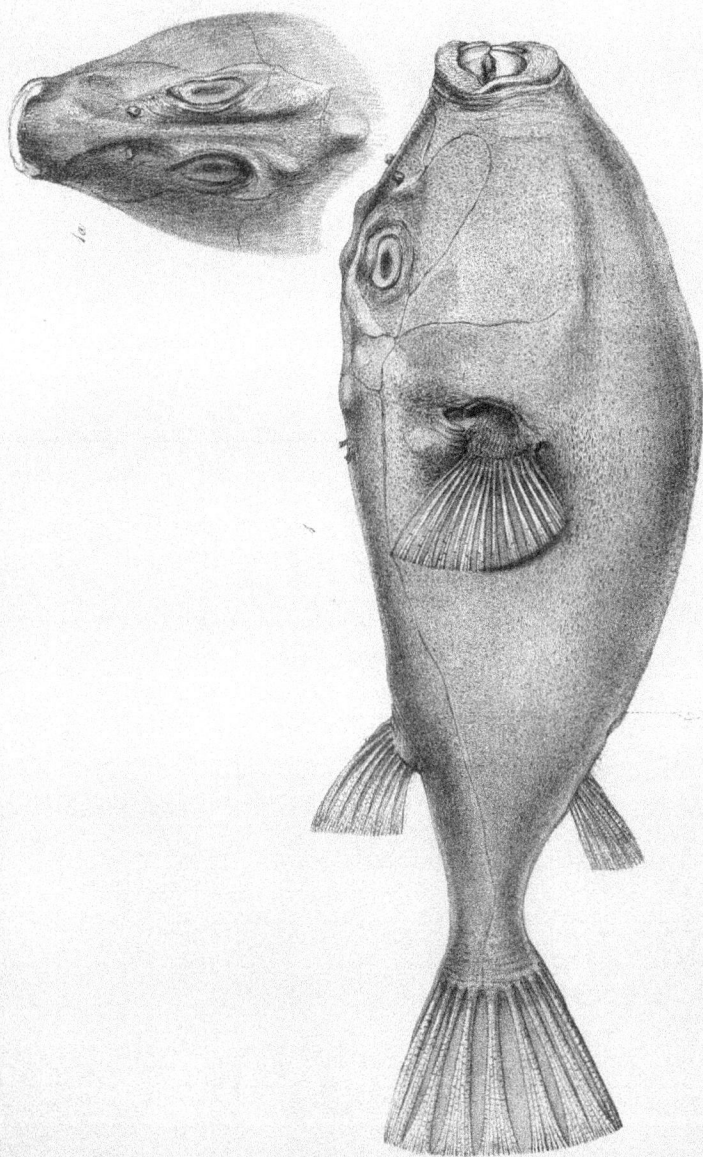

P. Sawerby del.

1. Tetrodon æquilabiatus
1 a. —— Dorsal View.
Nat. size.

FISH. 155

FAMILY.—BALISTIDÆ.

1. BALISTES VETULA. *Bl.*

Balistes Vetula, *Bloch,* Ichth. tab. 150.
———————— *Duperrey,* (Voyage) Zoologie, p. 114, pl. 9. fig. 2.

FORM.—Body deep, subrhombic, very much compressed; the greatest depth equalling half the entire length. Tail unarmed. Three or four larger scales than the others behind the branchial orifice. Pelvic bone projecting, prickly, connected with which is a fin consisting of about nine pairs of short rays. Above this fin, and parallel to its base, are two or three rows of short spines, but not much developed. First dorsal of three spines, commencing above the pectoral; first spine very strong and rough, the third not much smaller than the second. Second dorsal, and anal, which answer to each other, nearly even throughout, the anterior rays not being prolonged beyond the others. The caudal is injured, and its exact form cannot be determined. No lateral line.

D. 3—30 ; A. 27 ; C. 12 ; P. 14.

Length 1 inc. 10 lin.

COLOUR.—(*In spirits.*) Yellowish grey, becoming paler beneath. Three or four dark transverse streaks across the head from eye to eye: beneath the eye one or two indistinct streaks, passing off towards the branchial orifice: also two very distinct longer ones commencing on the upper part of the snout before the eyes, and passing obliquely across the cheeks towards the roots of the pectorals, parallel to those last mentioned. Besides the above, there are several obliquely transverse interrupted lines on the sides of the body: in one specimen, these lines are not well defined; in another, they are distinct, but so much interrupted as to have the appearance of spots arranged in a linear series. Two or three transverse lines encircling the tail; and some remains of longitudinal stripes on the second dorsal and anal fins.

The above description is that of two very small specimens of a species of *Balistes* taken by Mr. Darwin in Lat. 14° 20' South, Long. 38° 8' West, about sixty-five miles from land. I have very little doubt of their being the young of the *B. Vetula* of Bloch. The only respects in which they appear to differ from that species are the oblique lines on the back being carried completely across the sides in the form of lines of spots, and the anterior portions of the second dorsal and anal fins not being prolonged in a point ; but both these differences may be the effect of immaturity.

2. BALISTES ACULEATUS. *Bl.*

Balistes aculeatus, *Bloch,* Ichth. tab. 149.
———————— *Benn.* in Zool. of Beechey's Voy. p. 69. pl. 22. f. 2.

FORM.—Body deep, subrhombic. Tail armed with three rows of prickles, eleven in the uppermost row, about nine or ten in the middle one, and five or six in the lowermost. A few larger scales than the others behind the branchial orifice. Pelvic bone very rough and prickly, the spines that follow short, and not protruding much beyond the skin. First spine in the dorsal very strong, aculeated at the anterior edge, but not at the sides; no third spine in this fin. Second dorsal and anal even.   Caudal rounded.

D. 2—24; A. 21; C. 12; P. 13.

Length 2 inc. 3 lin.

COLOUR.—Not noticed in the recent state.   The ground colour has probably been altered by the spirit, but the markings are still very distinct, and accord tolerably with Bloch's figure, except that the oblique bands on the posterior part of the body, in front of and above the anal, are darker; while they alternate with four white ones, which are particularly conspicuous.   Possibly these white bands may have been originally blue, as the narrow stripes descending from the eyes to the pectorals, which evidently were of that colour, are nearly faded to a white. There is also a white stain on each side of the tail, where the spines are, which appears to have been blue originally: the spines themselves are deep shining black.

This specimen shows the black transverse bands between the eyes, and the broad band passing from the eye to the pectoral, between the narrow blue ones above alluded to, all represented by Bloch, but not observed by Mr. Bennett in the specimen figured in the " Zoology of Beechey's Voyage."

Habitat, Tahiti.

The above specimen was taken by Mr. Darwin at Tahiti.   It is quite small, and differs in some respects from the figures of Bloch and other authors, but it is evidently referable to the *B. aculeatus.*   The species is probably subject to variation in respect of colouring.

1. ALEUTERES MACULOSUS.   *Richards.*

Aleuteres maculosus, *Richards.* in Proceed. of Zool. Soc. 1840. p. 28.

FORM.—Oval, somewhat approaching to fusiform behind, very much compressed.   The greatest depth one-third of the entire length.   Skin covered with little granular points, terminating in very minute bristles, and communicating a slight roughness to the touch, when the finger is passed from tail to head.   Snout rather prominent and acute: jaws equal.   Dorsal spine springing from above the middle of the orbit of the eye; strong, with four rows of sharp prickles at the four angles, pointing downwards, and very regularly set: second dorsal spine very minute.   The second dorsal and anal fins have been lost in this specimen, and their form and number of rays cannot be determined.   The pectorals are small, each with twelve rays. Caudal rounded, also with twelve rays.

Length 5 inc. 4 lines.

COLOUR.—" Mottled with pale blackish green, leaving white spots."—D.—In its present state, the skin is nearly gone from long maceration in impure spirit: such portions as are left accord well with Dr. Richardson's description, appearing of a mouse-grey, with darker mottlings. There are three or four rather indistinct dark asciæ across the caudal.

Habitat, King George's Sound.

I have scarcely any doubt of this being the *A. maculosus* described by Dr. Richardson, in his recently published notes on a collection of fishes from Van Diemen's Land. Mr. Darwin's specimen, which is in bad condition, was obtained by him in King George's Sound.

2. ALEUTERES VELUTINUS. *Jen.*

A. pallide fuscescens, fasciis quatuor obscurioribus, longitudinalibus, indistinctis; pinnis pallide aurantiis: corpore oblongo-ovali elongato; cute delicate hispidâ, scabrâ: rostro producto, apice obtuso: spinâ dorsali aculeis lateralibus deflexis, uniseriatis: pinnis dorsali secundâ et anali multum ante caudalem desinentibus.

D. 2—33 ; A. 31 ; C. 12 ; P. 13 vel 14.

Long. unc. 8.

FORM.—Elongated, approaching to oblong-oval, the tail rather slender. Greatest depth exactly one-fourth of the entire length, and equalling the length of the head, this last being measured to the upper angle of the oblique branchial orifice. Back slightly arched, the curvature rather exceeding that of the belly. Profile in front of the dorsal spine falling very gradually, and not much out of the rectilineal. Snout considerably produced, but blunt at the extremity. Mouth small ; jaws equal ; teeth strong, and very sharp. Eyes round, placed exactly above the branchial orifice. The grains on the skin are coarser than in the *A. maculosus*, and the bristles springing from them longer and more developed, especially on the posterior part of the body, communicating a harsher feel to the touch : these bristles are slightly hooked at their extremities, the tips being turned towards the tail.

Dorsal spine strong, situate above the posterior part of the orbit, with only two principal rows of prickles, one on each of the two lateral edges ; anteriorly granulated at bottom, with a few rudimentary prickles towards the apex, but posteriorly almost quite smooth. Second spine very minute. The distance from the first spine to the commencement of the second dorsal fin equals twice the length of that spine. The anal commences under the fifth dorsal ray, and ends nearly in a line with the termination of that fin, but extends a trifle further. Both fins fall short of the caudal by a considerable space. Pectorals rather small. The caudal is worn at the end, but appears to have been either square or slightly rounded.

COLOUR.—"Very pale brown : fins pale orange."—D.

A *second specimen* is smaller than the above, measuring six inches and three-quarters in length. It is exactly similar in respect to form, and general colour ; but the sides are marked with four tolerably distinct longitudinal bands, extending from the branchial orifice to the caudal, rather darker than the ground on which they are traced. There is very little indication of these bands in the first specimen.

Habitat, King George's Sound.

This species was taken by Mr. Darwin in King George's Sound, and appears to be new. It has some points of resemblance with the *Balistes Ayraud* of Quoy and Gaimard, but in that the dorsal fin is said to extend to the caudal,[*] which is far from being the case here.   I have named it *velutinus*, in respect of the minute bristles which cover the skin, somewhat resembling the pile of velvet.

OSTRACION PUNCTATUS. *Schn.*

L'Ostracion pointillé, *Lacép.* Hist. Nat. des Poiss. tom. i. p. 455. pl. 21. fig. 1.

Ostracion punctatus, *Schneid.* Syst. Ichth. p. 501.

————Meleagris, *Shaw*, Nat. Misc. pl. 253.

This well-marked species of Ostracion, first described by Lacépède from Commerson's MSS., and afterwards figured by Shaw, in his "Naturalist's Miscellany," under the name of *O. Meleagris*, was obtained by Mr. Darwin at Tahiti, where it had been previously observed by Captain Cook.

There are two specimens in the collection, both exactly similar, and of the same size, measuring a trifle more than three inches and a half in length.   They also accord well with Shaw's figure.   Lacépède, in his description, speaks of the anal fin as being more extended than the dorsal, and as having eleven rays ; but in both Mr. Darwin's specimens, I find the number of rays in these two fins the same.   The formula is as follows :

D. 9 ; A. 9 ; C. 8 ; P. 10.

Schneider has noticed this species twice ; first under the name of *lentiginosus*, and again under that of *punctatus*.

---

[*] This character, though mentioned in the description, is not, however, represented in the figure. See *Freycinet's Voyage (Zoologie)*, pl. 47. f. 2.

CYCLOSTOMI.

FAMILY.—PETROMYZONIDÆ.

MYXINE AUSTRALIS. *Jen.*

FORM.—Scarcely differing from the *M. glutinosa,* but apparently rather more slender in proportion to its length. Mouth and cirriform appendages the same. Branchial orifices two, very near together, placed beneath, at a little beyond one-fourth of the entire length. A very conspicuous row of pores along each side of the abdomen. The tail seems somewhat sharper than in the *M. glutinosa,* and the rays of the low fin which turns round its extremity rather more distinct. Vent distant from the end of the tail rather less than one-eighth of the entire length.

Length 11 inc. 6 lin.

COLOUR.—" Above coloured like an earth-worm, but more leaden; beneath yellowish; head purplish."—D.

Habitat, Tierra del Fuego.

Mr. Darwin obtained this species by hook amongst the kelp, in Goree Sound, and other parts of Tierra del Fuego, where he observes it is abundant amongst the rocky islets. Its extreme southern locality would suggest the idea of its being distinct from the *M. glutinosa* of the northern seas; yet the differences between the two, upon comparison, are very slight, and, if it really be so, as I have ventured to consider it, it requires an examination of more specimens to lay down its exact specific character.

Mr. Darwin has made some interesting remarks on the habits of this fish. He observes that it is " very vivacious, and retained its life for a long time; that it had great powers of twisting itself, and could swim tail first. When irritated, it struck at any object with its teeth; and by protruding them, in its manner, much resembled an adder striking with its fangs. It vomited up a *Sipunculus* when caught." He adds, that he " observed a milky fluid transuding through the row of lateral pores."

APPENDIX.

THE following Appendix contains descriptions of a few species, which were omitted to be noticed in their proper places; and further remarks with respect to some, which will be found in the body of the work.

FAMILY.—PERCIDÆ.

1. APHRITIS UNDULATUS. *Jen.*

PLATE XXIX. fig. 1.

A. elongatus: lateribus supra pallide olivaceis, fasciis transversis abbreviatis, lineisque longitudinaliter undantibus, nigris; lateribus infra argenteis: pinnis dorsalibus et caudali punctatis; pinnis, reliquis, et lineâ laterali, albidis.

B. 6; D. 8—25; A. 1/22; C. 14, et 6 brevioribus; P. 22; V. 1/5.

LONG. unc. 3. lin. 1.

FORM.—Elongated; the depth about one-sixth of the entire length; the thickness two-thirds of the depth. Head four-and-a-half times in the length. Profile falling very gradually at first, but more rapidly in advance of the eyes, causing the snout to appear rather obtuse. Mouth small: maxillary slender, hardly reaching to a vertical line from the anterior margin of the orbit: upper jaw slightly longer than the lower, and very protractile. Teeth very minute, forming a narrow velutine band: a patch on the chevron of the vomer scarcely visible, but capable of being very distinctly felt; none apparent on the palatines. Eye one-fourth the length of the head, and distant one diameter from the end of the snout: the interocular space rather less than the diameter. Snout slightly indented, or furrowed out in front of the eyes. A series of impressions on the lower jaw, and along the limb of the preopercle, but much less obvious than in the next species, and not distinctly porous. Preopercle with the ascending margin vertical, the angle at bottom rounded; the limb broad and distinctly marked, with the boundary line between it and the cheek slightly elevated into a ridge. The opercle, with its membrane, produced backwards in an angle, the subopercle being visible beneath. The branchial membrane six-rayed, and fastened to the isthmus underneath, the aperture commencing beneath the ascending margin of the preopercle.

Lateral line commencing at the upper angle of the gill-opening, and following the curvature of the back at one-fourth of the depth, and preserving this direction throughout its course, not

2

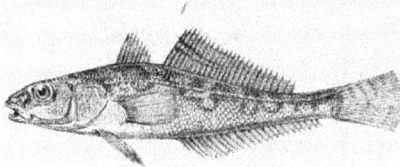

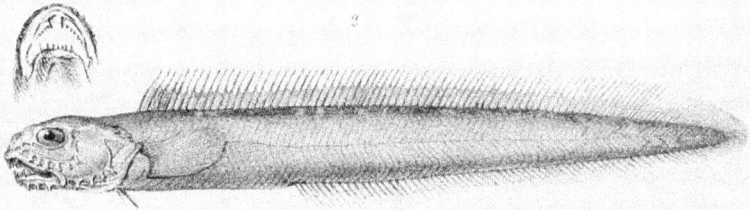

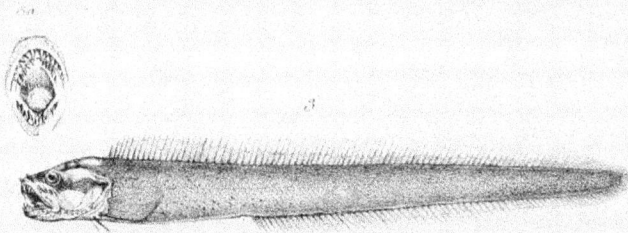

Waterhouse Hawkins del.

1 Aphritis undulatus
2 Iluocates fimbriatus
2a   Magnified View of Teeth   Nat. Size
3 Phucocates latitans.
3a   .   .   Teeth.

falling to the middle before losing itself in the caudal. Scales small, covering the whole head and body, except the snout in front of the eyes, the jaws, and the limb of the preopercle. The free portion of each scale marked with several small concentric circles, the free edge finely ciliated: the basal portion with a fan of seven striæ, and the spaces between these deeper striæ with minuter striæ running transversely: the basal margin cut square.

Pectorals attached rather low down, and a little posterior to the terminating angle of the opercle; their length about three-fourths that of the head: the fourth to the eighth rays longest; the first ray only half the length of the second; the first two, and the last three or four, simple; the rest branched. Ventrals about four-fifths the length of the pectorals, and in advance of those fins by nearly half their own length; their spine very distinct. First dorsal short, commencing immediately above the insertion of the pectoral: all the spines very slender, with the intervening membrane delicate; the second longest, equalling about half the depth; the third and following ones gradually decreasing. Second dorsal long, separated from the first by a very small interval, and occupying a space just equal to the distance between its commencement and the end of the snout: the rays gradually decreasing in length from the anterior ones, which equal three-fourths of the depth; all simple, or if branched, only so at their extreme tips. The interval between the second dorsal and the caudal contained eight-and-a-half times in the entire length. Anal commencing under the sixth ray of the second dorsal, or exactly at the middle point of the entire length, caudal excluded; extending a trifle beyond the second dorsal, but in other respects answering to that fin. Caudal square when spread, but very slightly notched when the rays are close; contained six-and-a half times in the entire length; the principal rays branched.

COLOUR.—(*In spirits.*) Back and upper half of the sides pale olivaceous, with about seven or eight abbreviated, transverse, dusky fasciæ; beneath these are two irregular lines undulating longitudinally in a zig-zag manner, and having rather a tendency to meet at the angles, so as to form a connected longitudinal chain of diamond-shaped links. Lower portion of the sides and abdomen silvery. Tubal pores of the lateral line white, making this line very evident. Dorsal and caudal fins speckled with small dusky spots and points. Pectorals, ventrals, and anal, quite plain, and whitish.

A *second specimen* in the collection exactly resembles the above, except in being not quite so large, and in having a ray less in each of the two dorsal fins.

Habitat, Chonos Archipelago, W. coast of S. America.

The genus *Aphritis* was first established by M. Valenciennes, in the appendix to the eighth volume of the " Histoire des Poissons," for the reception of a small Percoid fish obtained by MM. Quoy and Gaimard in Van Dieman's Land, inhabiting fresh-water. The species above described, which was taken by Mr. Darwin in Lowe's Harbour, South of Chiloe, appears to be referable to the same genus. It differs, however, in many respects from the *A. Urvillii,* the only one which Valenciennes has noticed. The relative situation of the first dorsal with respect to the pectorals, and of the anal with respect to the second dorsal, is different: there are fewer rays in the anal, and more in the second dorsal: the upper, instead of the lower jaw, as represented in Valenciennes's figure, is

Y

rather the longest; neither can I discern any teeth on the palatines, though there is a patch of very minute ones in front of the vomer.

That this species really belongs to *Aphritis*, would seem indicated not merely by the aggregate of its external characters, but by the internal structure also, which was examined in one of the two specimens brought home by Mr. Darwin, and found conformable to what is stated by Valenciennes, in this respect, of the *A. Urvillii*. The stomach is large, with four very distinct cœcal appendages, and there is no air-bladder.

The *A. undulatus*, which I have so named in reference to the undulating longitudinal lines on the sides, is very Cypriniform in general appearance, and not altogether unlike the common minnow, *Cyprinus Phoxinus*.

2. APHRITIS POROSUS. *Jen.*

A. brevior : pallide olivaceus, lateribus fasciis transversis obsoletis nigricantibus ; pinnis omnibus brunneis : maxillâ inferiore, et limbo preoperculi, poris conspicuis circiter novem, suborbitalibus circiter quinque, seriatim dispositis.

D. 8—25 ; A. 1/22; C. 14, &c. ; P. 23 ; V. 1/5.

LONG. unc. 2. lin. 5.

FORM.—Not so much elongated as the last species: the depth rather more than one-sixth of the entire length, and the head only four times in the same. Also distinguished by a row of large mucous pores on the lower jaw, passing upwards posteriorly, and continued along the limb of the preopercle : the number of these pores on each side is nine or ten : a row of similar pores, amounting to about five, passes backwards from a little above the end of the maxillary beneath each eye. In other respects, the form is similar to that of the last species, excepting that the interval between the second dorsal and the caudal is only one-eighth of the entire length, in consequence of the body being less elongated.

COLOUR.—(*In spirits*). Back and upper half of the sides, olivaceous brown; beneath silvery. No very obvious markings; but traces may be seen of six or seven transverse dusky fasciæ, reaching from the back to a little beneath the lateral line, which were probably more distinct in the recent state. All the fins brownish; the dorsal a little powdered with dusky specks. The fourth and fifth rays of the ventrals are white, and appear to have been always of a different colour from the rest of the fin.

Habitat, Coast of Patagonia.

This appears to be another new species of *Aphritis*, taken by Mr. Darwin on mud-banks, in Port Desire, central Patagonia. It is very closely allied to the *A. undulatus*, but, I conceive, certainly distinct. There is but one specimen in the collection.

FAMILY.—SCORPÆNIDÆ.

APISTUS ———— ?

Mr. Darwin's collection contains a species of this genus procured in King George's Sound, New Holland, which, from the bad state of preservation of the specimen, it is scarcely possible to identify with certainty. Possibly it may be new, as it does not seem to accord very exactly with any of those described in the "Histoire des Poissons;" but I shall not consider it such, nor do more than point out a few of its more obvious characters.

It is not determinable, whether it was originally one of the naked species of this genus, or whether the scales have been rubbed off, but probably the former. The suborbital and preopercular spines are strong, and considerably developed: the former reaches back further than the maxillary, and nearly to the posterior part of the orbit, and has another very small spine at its base. The lower jaw advances beyond the upper. The head is about one-third of the entire length. The eyes are large, their diameter being contained about three and a half times in the length of the head. The dorsal commences in a line with the ascending margin of the preopercle. The first spine is half the length of the second; the second is a little shorter than the third, which is longest, and equals two-thirds of the depth of the body; the fourth and succeeding ones decrease very gradually; the soft portion of this fin is a little higher than the hinder part of the spinous. The first anal spine is rather more than half the length of the second, which is the strongest of the three, though not much longer than the third. The pectorals are rather pointed, and a little shorter than the head. The ventrals are attached a little behind the pectorals, and are not very much shorter than those fins.

The following is the fin-ray formula:

D. 13,9; A. 3/6; C. 11, &c.; P. 11; V. 1/5.

Length 4 inches 6 lines.

The species to which this approaches nearest would seem to be the *A. niger* of Cuvier and Valenciennes; but there is no appearance of the small elevations on the skin resembling hairs, which those authors mention in their description of this last, and, on the whole, I am inclined to consider it as distinct.

AGRIOPUS HISPIDUS. p. 38.

Notwithstanding what I have advanced in regard to this species, further consideration has inclined me to suspect, that it may prove ultimately only the young of the *A. Peruvianus*. In that case, however, it would appear that the absence of vomerine teeth can only be assigned as a character of this genus in the adult state.

FAMILY.—SCIÆNIDÆ.

OTOLITHUS ANALIS. *Jen.*

This new species of *Otolithus* is from Callao : it was omitted to be noticed in the body of the work. There is but one specimen in the collection, in bad condition, and not admitting of a very detailed description ; but it is evidently distinct from all the species described by Cuvier and Valenciennes.

It is rather more elongated than the *O. Guatucupa*, the depth being not much more than one-fifth of the entire length. The head is long, and contained three and a half times in the same. The lower jaw is considerably the longest. The teeth above are small, and sharp-pointed, apparently in two rows, the outer row being a little stronger than the inner: there are two very strong canines in front, springing from between the rows. Below, the teeth are in two rows in front, and one at the sides ; those in front small, but those at the sides unequally sized, three or four, standing at intervals, being much stronger than the others, and very sharp. Diameter of the eye about one-sixth the length of the head ; its distance from the end of the snout one diameter and a half.

The lateral line is continued to the extremity of the caudal, between the ninth and tenth rays. There is a small interval between the two dorsal fins. The second dorsal, as well as the anal, are longer than in the *O. Guatucupa*, with more soft rays, especially the anal. The pectorals are narrow and pointed, and between one-half and two-thirds the length of the head. The ventrals are attached almost exactly beneath the pectorals. The caudal appears to have been square.

D. 9—1/24 ; A. 1/16 ; P. 17 ; V. 1/5.

Length 12 inches.

COLOUR.—The colours were not noticed when recent, and can hardly be judged of now. The general tint appears to have been silvery. If there were any markings, no traces of them remain.

Habitat, Callao, Peru.

This species has a longer anal than any of the American species described by Cuvier and Valenciennes. I have in consequence named it *analis*.

PRIONODES FASCIATUS. p. 47.

It has been suggested that this may be nothing more than a monstrosity. Whether this be really the fact or not, can only be determined by the examination of more specimens. But in either case, I am so satisfied now of its being a *Serranus* in all its essential characters, that I conceive it never can be placed in a different family from that genus. If the fact be established of its never possessing vomerine and palatine teeth, such a character can, at the very utmost, serve only to distinguish it as a subgenus in that group. But every day is bringing more and more to light the small value of that character.

STEGASTES IMBRICATUS. p. 63.

I am informed by Mr. Lowe, of Madeira, that this is the same as the *Glyphisodon luridus* of Cuvier and Valenciennes.* Their description is so short, that I failed to recognize it ; and I was induced to consider it as a new genus from the circumstance of its possessing vomerine teeth. Whether these teeth exist in any other species, or have only been presumed absent in all, because not found in some, I am not aware. But here again we see how little such a character is to be depended upon.

FAMILY.—BLENNIDÆ.

Mr. Darwin's collection contains two new forms from South America, closely allied to each other, yet forming distinct genera, and which will not enter into any of those described by authors. At first it was conceived that they were Malacopterygian fishes, more especially from their having all the rays in the dorsal and anal fins articulated ; and to belong to the Apodal division of that group, from their being supposed to be without ventrals ; but, on a closer inspection, the ventrals, which are very small, were found to have been overlooked, and it was evident altogether that the true place of these fishes in the system was amongst the *Blennidæ*. The mention of this circumstance will explain why they were omitted to be noticed in their proper place.

These two genera, so far as can be judged from the situations in which Mr. Darwin obtained them, have the same habits as the Blennies, lurking under stones and weeds ; and I propose to give them respectively the names of *Iluocœtes* and *Phucocœtes*.

GENUS.—ILUOCŒTES.* *Jen.*

Corpus elongatum, antice subcylindricum, postice compressum, ensiforme, læve, nudum, alepidotum. Rostrum breve, obtusum, rotundatum, ultrà maxillam inferiorem productum. Dentes acuti, subconici, in utráque maxillá uniseriati : supra canini duo fortes, curvati, antici, et præ serie exstantes : in vomere dentes pauci acuti aggregati ; in utroque palatino uniseriati. Lingua lævis. Oculi grandes, prominuli. Apertura branchialis mediocriter fissa, membraná quinque-radiatá. Maxillæ, os suborbitale, et præoperculum, tubiporis cutaneis brevibus ad margines fimbriatæ. Pinnæ ventrales jugulares, minutæ, gracillimæ, triradiatæ. Pinnæ dorsalis et analis prælongæ, caudali coalescentes, radiis omnibus articulatis.

If I am right in placing this new genus amongst the *Blennidæ*, it will evidently take its place next to *Zoarces*, to which it is more nearly allied than to any other

* *Hist. des Poiss.* tom. 5. p. 356.　　　† Ab ιλυς limus, et κοιτη cubile.

group in that family.   It agrees especially with *Zoarces*, not only in general form, but in having all the dorsal and anal rays articulated, (excepting one in the dorsal, which possibly may be an accident in the only specimen examined,) and in having the ventral fins extremely small.   On the other hand, it departs from that genus, in having the body entirely naked, and free from scales ; in the two remarkable canines in front of the upper jaw, and in having teeth on the palate ; also in having no notch at the posterior part of the dorsal.   It is further remarkable for its large prominent eyes, and the rows of tubipores on the cheeks.   Amongst the true Malacopterygians, it approaches nearest to *Ophidium*, and, but for the circumstance of its possessing ventrals, it might perhaps be ranged under that genus.   It is, however, evidently a connecting link between the *Apodal Malacopterygians* and the *Blennidæ*.

Like the Blennies, this genus has neither cœcal appendages nor air-bladder. The intestinal canal is rather ample, with a few coils, but of tolerably equal dimensions throughout.

There is but one species of this new genus in the collection, which is from the Archipelago of Chiloe.   The detailed description of it is as follows :—

ILUOCŒTES FIMBRIATUS.   *Jen.*

PLATE XXIX. Fig. 2.

FORM.—Very much elongated, subcylindric anteriorly, compressed and ensiform behind.   Greatest depth in the region of the pectorals, and about one-tenth of the entire length.   Head, measured to the extreme point of the gill-cover, five and a half times in the same.   The head is rather larger than any part of the body, its depth and thickness being equal, and each a trifle less than the depth of the body.   Crown and forehead a little flattened, whence the profile descends in a curve before the eyes.   Snout blunt and rounded, projecting, both in front and at the sides, beyond the lower jaw.   Gape wide, and reaching to beneath the middle of the eye.   Intermaxillary somewhat protractile at the sides, but not in front ; shorter than the maxillary, with a considerable intervention of membrane between the extremities of the two bones, which are not united posteriorly, excepting by the membrane just mentioned.   Maxillary long, rather slender, of nearly uniform breadth and thickness throughout, retiring in part beneath the suborbital, and reaching backwards to a vertical from the posterior part of the orbit.   Intermaxillary with a single row of small, pointed, subconical, slightly curved teeth : in front of these, and quite at the anterior extremity of the jaw, two strong, hooked, regular canines :* the teeth in the row rather wide asunder, and set a little irregularly, about thirty in number.   In the lower jaw, teeth few in number, scarcely more than eight or ten in front, rather stronger than the intermaxillary series, followed by a moderate canine on each side, these last smaller than the ones above : at the sides of the lower jaw, beyond the canines, scarcely more than one or two small teeth (possibly others are fallen).   A small cluster of three or four teeth on the fore part of the vomer, like those in front of the lower jaw, and a

* One of these is gone in this specimen, but the socket in which it was implanted is obvious.

row on each palatine. Pharynx also armed with strong teeth; but the tongue, which is free at the tip, and rounded, smooth. Eyes large and prominent, and elevated rather above the line of the profile: their diameter one-fourth the length of the head; their distance from the end of the snout one diameter; the interocular space reduced to a narrow channel, and scarcely equalling half a diameter.

Opercle of a triangular form; subopercle lanceolate, projecting further than the opercle, and passing upwards and backwards to form the terminating angle of the gill-cover. Gill-opening of very moderate extent; the branchial membrane fastened down underneath, with five rays. Skin smooth and naked, loose, and probably very mucous in the living fish. Apparently no lateral line. The edges of both jaws curiously fringed each with a row of tubipores, or cutaneous appendages in the form of tubes, having pores at their terminal extremities for the exudation of mucus. The row on the upper jaw is continued along the margin of the suborbital on to the cheek; that on the lower is carried upwards to form an edging to the preopercle. There is also one of these tubipores at each nostril, another behind each eye, and a third on each side of the nape.

The dorsal commences above the terminating angle of the gill-cover, and extends the whole length of the body: its height nearly uniform throughout, equalling half the depth: the rays slender; all articulated, except the third, which is spinous, and shorter than those which precede and follow it; mostly simple, but some of the posterior ones slightly divided at their tips. Vent situate beneath the termination of the first quarter of the dorsal. The anal begins immediately behind it, and, like the dorsal, is carried on to the end of the body, to unite with it in forming a pointed caudal; all the rays soft and delicate. Pectorals rather pointed, with the middle rays longest, and about two-thirds the length of the head. Ventrals very minute and narrow, of only three rays, and appearing like one filament, about one-third the length of the pectorals; attached in front of the pectorals, and nearly in a line with the gill-opening.

B. 5; D. about 80; A. about 60; C. about 15; P. 16; V. 3.

Length 5 inc. 9 lin.

COLOUR.—Not noticed in the recent state. In spirits it is nearly colourless, with the exception of a dark bluish line along the base of the dorsal; upper part of the head, and nape, also stained with the same dark tint.

Habitat, Archipelago of Chiloe.

This species was taken by Mr. Darwin under stones. There is but one specimen in the collection, and it would be very desirable to see others, in order to ascertain whether the circumstance of the *third* dorsal ray *alone* being spinous, (those that precede as well as follow being articulated), is merely accidental in the one above described, or really characteristic of the species. If the latter, it is an anomaly,—a single spine thus occurring in the middle of a soft fin,—of which I know no other example.

Genus.—PHUCOCŒTES.* *Jen.*

Corpus elongatum, compressum, nudum, alepidotum, porosissimum. Rostrum breve, obtusum. Dentes acuti, subconici, in maxillá superiore uniseriati, in inferiore bi-vel tri-seriati; supra canini duo fortiores, antici, et præ serie exstantes: in vomere dentes duo vel tres acuti, quorum unus fortis ; in utroque palatino uniseriati. Lingua lævis. Oculi parvi. Apertura branchialis arctissima, membraná sex-radiatá. Maxillarum margines poris conspicuis longitudinaliter dispositis, simplicibus, haud in tubos productis. Pinnæ ventrales, dorsalis et analis, ut in genere præcedenti.

This genus differs from *Iluocœtes*, in having the head and eyes smaller, the snout scarcely at all produced, the teeth in front of the lower jaw in two or three rows, and especially in the contracted gill-opening, which is reduced to a small hole, relatively not much larger than in the Eels, and in the branchial membrane having six rays. It wants also the tubal cutaneous appendages on the jaws and cheeks, in the place of which are rows of simple pores. It may be added that the whole skin is every where studded with pores; smaller, however, than those which form the maxillary series. The fins are similar, including the minute ventrals; but the tail and caudal are more rounded, and the membrane investing the rays of the dorsal and anal is more fleshy, so as hardly to allow of the rays being counted.

This genus is yet more eel-like, and more malacopterygian in general appearance than the last, serving to make the passage from the *Blennidæ* to the apodal division of the soft-finned fishes, still more gradual and evident. Mr. Darwin obtained it at the Falkland Islands. There is but one species in the collection referable to it.

PHUCOCŒTES LATITANS. *Jen.*

PLATE XXIX. Fig. 3.

FORM.—Still more elongated than the *Iluocœtes fimbriatus*, but not tapering so much to a point posteriorly, the tail being blunter and more rounded. Greatest depth about one-eleventh of the entire length: head one-seventh. Head more compressed, its thickness being only three-fourths of its depth. Nape rather more elevated, and the profile more sloping, its descent commencing at a more backward point. Snout equally short and rounded, but much less projecting over the lower jaw. Gape, intermaxillary, and maxillary, similar: also the teeth; only the pair of canines above, standing in front of the series, are smaller; and below, the teeth in front are in two or more rows. On the fore part of the vomer is one strong tooth, and apparently one or two other smaller teeth with it; on each palatine a row, one or two of the anterior ones being stronger than the others. Eyes very much smaller; their diameter scarcely more than one-seventh the length of the head; not sufficiently high in the cheeks to cut the line of the profile: interocular space slightly convex.

* A φυκος fucus, et κοιτη cubile.

Pieces of the gill-cover on the whole similar, but the branchial aperture much smaller, the fissure not descending below the level of the upper part of the pectoral: branchial membrane with six rays. Skin quite naked, and thickly studded all over with mucous pores. Also some very large and conspicuous pores in rows on the jaws and cheeks, but not elevated into cutaneous tubes, excepting the nostrils, which are tubular.

Dorsal and anal with all the rays articulated, and the greater part of them simple, but some toward the tail a little branched at their tips. Caudal not so pointed as in *Ruocœtes*. Pectorals and ventrals similar, but the latter a trifle longer and broader in proportion.

Length 4 inc. 7 lines.

Colour.—(*In spirits.*)—Brown, with the jaws, under part of the head, and lower half of the cheeks, whitish ; also a whitish fascia extending longitudinally from behind each eye to the upper angle of the opercle.

. *A second specimen* in the collection is smaller than the above, measuring only two inches and a half in length. It is in bad condition, but does not appear to differ, except in having the anterior canines above but very little developed.

Habitat, Falkland Islands.

Both individuals of this species were taken by Mr. Darwin in the Falkland Islands. " Caught amongst kelp."—D.

ERRATA.

Page 6, line 2, from the bottom, for *versus apicem* read *apicem versus*.
   9, — 7, ——————— for *versum angulum* read *angulum versus*
   13, — 4, ——————.— for *ciliatis* read *ciliatæ*
   18, — 17, from the top, for *duobus* read *duabus*.

INDEX.

4

www.ingramcontent.com/pod-product-compliance
Lightning Source LLC
Chambersburg PA
CBHW031428270326
41930CB00007B/616

*9 7 8 1 5 2 8 7 1 2 1 1 8 *